Amazing Sun Literacy

Science Exploration by Rolf A. F. Witzsche

© Text Copyright Rolf A. F. Witzsche 2018
all rights reserved

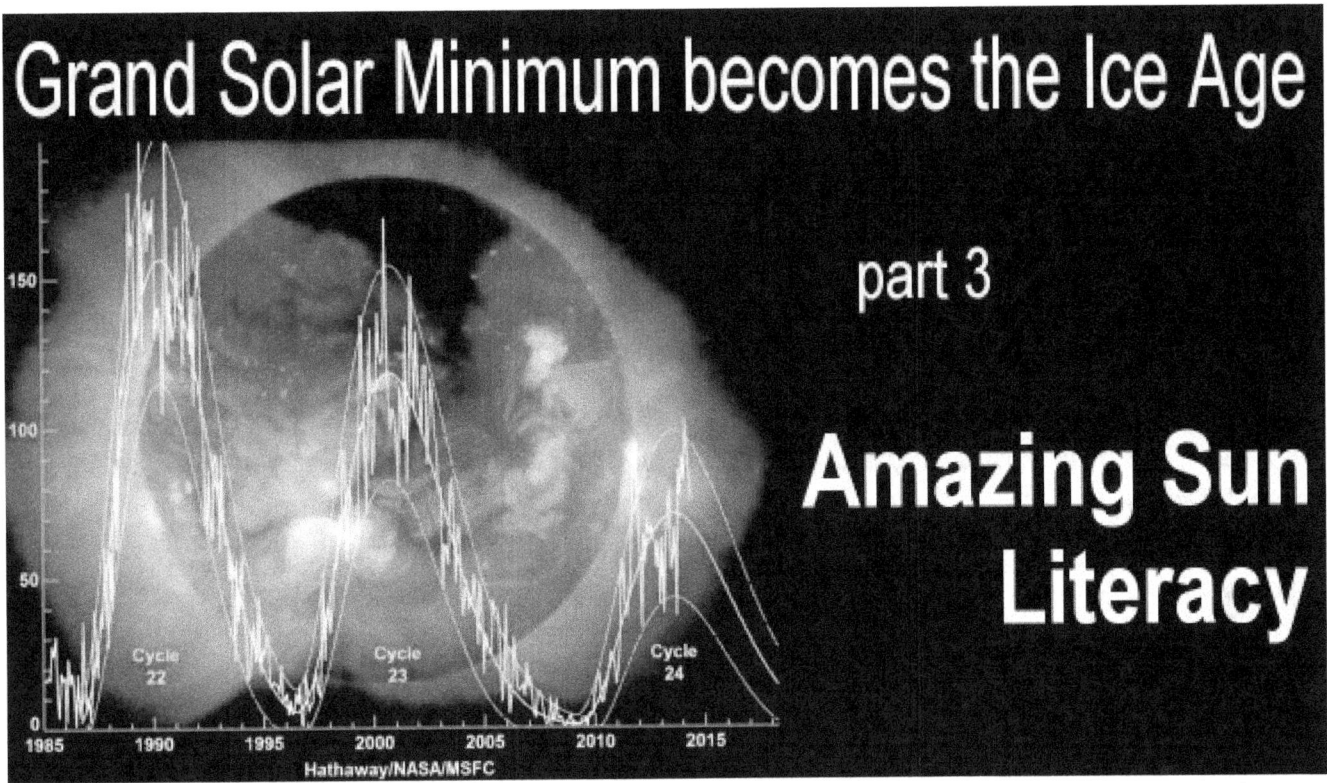

The book contains the transcript and images of the science exploration video by Rolf A. F. Witzsche, with the above title at: http://www.ice-age-ahead-iaa.ca-- The book is a part of the transcripts series.

Lead In:

The magnetic field of the Sun's north pole has vanished as another shoe drops on the 'rush' to the next Ice Age with its potential phase shift in the 2050s.

The magnetic polarity reversal at the polar regions of the Sun, which happened in 2013, didn't materialize for the North Pole of the Sun. The Sun, presently, does not have a magnetic north pole anymore. The northern magnetic polarity is neutral.

How is this possible? Let's ask the Sun what it has to say about that. For this, we need to decipher its language of the principles that make it tick.

As we let the Sun 'speak', it tells us an amazing story, of why its solar cycles are getting longer, its polar magnetic reversals weaker, and its sunspot numbers are getting smaller. It also tells us how this is connected with the fact that it rotates its surface faster at the equator than at its poles.

The opening music for the original video is the opening music performed at the 2017 BRICS Summit gala event. The up-beat music reflects our human potential to build us a new world that the Ice Age cannot touch, and to build with it a brighter and richer civilization than we presently have.

The Ice Age imperative that we have gleamed from the diminishing Sun may be one of the biggest drivers that we have to inspire us to work together as a single humanity, to create a New World for us, thus to lay aside the small-minded terrors that presently darken our existence and our future, such as the nuclear weapons threats, imperial wars, economic collapse, political insanity, and so on.

Yes, we have the potential to build us a bright future as we discover evermore of the already near Ice Age that is poised to grip our world. The more we allow the Sun to present its case, we will find us inspired to move with it, not in ignorance of it. And why shouldn't we move with it? The principles of the solar system lay before us for the grasping. Our response then follows naturally.

Universal good, as a concept, is already on the agenda in China with the Belt and Road initiative. It is also on the agenda in the form of the call by Russia for a new world-security architecture that can actually work. The 'rush' for meeting the common aims of mankind appears to have begun, and its beginning appears to be on track with the astrophysical 'rush' to the next Ice Age that we cannot evade but can raise ourselves above with building us a New World that the Ice Age dynamics cannot touch.

So it is, that what seemed hopeless just a few years ago, is amazingly now sweeping the world, sweeping it clean of archaic myths and modes, sweeping in truthful perceptions from numerous directions at once, even to the point of letting the Sun speak for itself, thus building us up into the direction that is raising evermore people out of their easy chair, into action.

This exploration presented here is a part of the unfolding movement towards the inevitable universal good.

Contents

- ➤ The Sun's northern magnetic field has vanished 10
- The Sun's polar magnetic fields 11
- As the Sun is understood in mainstream science 12
- Putin had something to do with this 13
- We could also ask the Sun 14
- A 19-page research paper on the subject 15
- To look under its skin 16
- Can we learn the language of the Sun? 17
- Advanced deciphering of the 'language' of the Sun 18
- Ignoring what the Sun has to tell us 19
- ➤ On the development of languages 20
- Development of written languages 21
- Some languages have become lost 22
- Epic literary works attributed to Homer 23
- Language development by Dante Alighieri 24
- Cosmic universe has a specific 'language' 25
- ➤ Universe 'speaks' the 'language' of plasma 26
- At the Los Alamos National Laboratory 27
- Plasma, simply is everywhere 28
- David Bohm, as his successor 29
- Called protons and electrons 30
- Electrons are the smallest components 31
- Cosmic space is filled with examples 32
- ➤ The plasma stream 33
- It speaks the language of magnetism 34

Termed the Lorenz force	35
Wires in cosmic space	36
As the plasma streams flow	37
The amazing plasma-flow geometry	38
He termed, the Primer Fields	39
Enormously large magnetic forces	40
➢ The plasma sphere	41
Plasma is not visible	42
When large volumes of plasma form a sphere	43
Such a sphere of plasma is a Sun	44
This 'electrically charged' sphere	45
The Sun is dark inside	46
The Sun comes to light as a star	47
The plasma Sun is located at the center	48
Primer fields develop at a node point	49
➢ Let's decode the 'language' of the Sun	50
Sun rotates significantly faster at its equator	51
➢ The puzzling differential rotation	52
If the Sun was a body of atomic elements or gases	53
A 40% difference in rotational speed	54
Rotation is driven from outside the Sun	55
The Sun located within a cylinder	56
By electromagnetic coupling	57
The Equator of the Sun rotates the fastest	58
Perplexing as mechanistic phenomena	59
The 'language' of the real solar dynamics	60
The 'language' of the plasma dynamics	61

The same applies to the Sun in visible light	62
Differential in solar activity	63
Sunspots remain largely a paradox	64
The Sun's 'active' region	65
The Sun's symmetric magnetic-field orientation	66
➢ The Sun's magnetic field	67
Opposite directions, symmetric to the equator	68
heliospheric current sheet	69
The Rings of Saturn, are at the planet's ecliptic	70
In comparison, the heliospheric current sheet	71
The perfect alignment	72
To serve as a guide for magnetic fields	73
Sunspots are damaged areas	74
The eruption of sunspots creates loops	75
The eruption flow, and the re-entry flow	76
Let's look at sunspot 2700	77
Another example from May 24/2015	78
The differently oriented polarization proves	79
Proof that the Sun's magnetic field is imposed	80
Loop structures that don't produce sunspots	81
Loops associated with the magnetic spots	82
Plasma loops are occur horizontally	83
Sunspots from the previous solar cycle	84
Because the Sun's magnetic polarity flips	85
The symmetric magnetic field reversal proves	86
➢ The 22-year cycle	87
How is the polarization reversal generated?	88

The answer is amazing	89
Differential creates a magnetic-field polarity	90
Magnetic reversal in the polar regions of the Sun	91
In plasma language the delay is recognized	92
The Sun's distance to the primer fields	93
For most of the Sun, the polarity reversal is not delayed	94
The reason why the solar activity peaks	95
The shock in the dome starts the magnetic field reversal	96
The primer fields being magnetically connected	98
A 2-way magnetic oscillation cycle	99
The solar system as a whole is diminishing	100
A 19-page research report	101
Increase of the duration of the solar cycle 24	102
➢ The vanished northern magnetic field	103
The polar magnetic field reversal in cycle 23	104
Polar polarity reversal didn't happen	105
The missing northern polar polarity	106
The solar system is crashing	107
Cycles 21 and 22, gave us a yardstick	108
➢ When the 'shoe' drops	109
Solar system in a state of transition	110
A kind of hibernation state	111
The resulting 70% colder Sun	112
A near total loss of agriculture across the world	113
The slowing solar heartbeat	114
➢ Interglacial climate cooling is now accelerating	115
The big warming spikes began to diminish	116

Solar minimum events became colder 117

The solar system is rushing back 118

The diminishing solar-wind pressure 119

Just another 'shoe' dropping off along the path 120

Science should be the voice of truth 121

The Ice Age, is a long-term 'solar cycle' 122

➤ The principle of universal good 123

To build us a new world in the tropics 124

Impossible in the landscape of zero-sum economics 125

Losers are balanced by the winner 126

Increase the productive capacity 100,000 fold 127

An atom is 100,000-time larger than its parts 128

The question will then be, what else can we build 129

She says 'Good' IS 'God' (universal) 130

On the platform of universal good 131

The lack of love for our humanity is still extensive 132

In the current arena of tenacious insanity 133

In the nuclear weapons trap for 70 years 134

Whereby it has the potential to change us 135

Why should we fail on this front? 136

➤ More from the author: 137

14 Libraries of books and video productions 137

Start of the exploration

➤ The Sun's northern magnetic field has vanished

http://www.zam.fme.vutbr.cz/~druck/Eclipse/ - an example of the amazing solar eclipse photography of Milloslav Druckmueller

An amazingly tragic catastrophe has happened to the Sun. The Sun has lost its magnetic polarity in the northern polar region. While the loss is not visually apparent by looking at the Sun with standard telescopes, even during eclipse conditions, the fact is that the Sun's northern magnetic field has vanished as if it never existed.

The Sun's polar magnetic fields

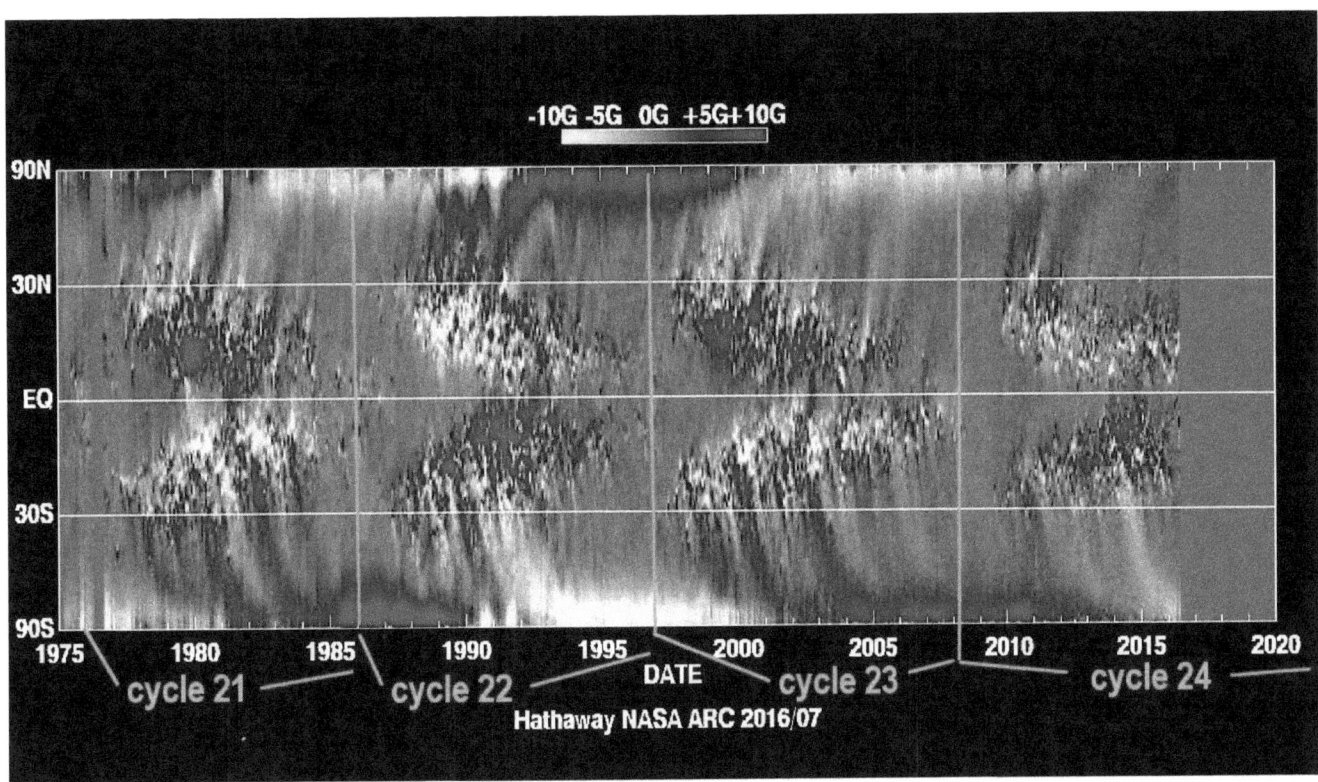

For as long as scientists have tracked the magnetic polarity of the Sun, which became possible from the 1970s on with specialized equipment, the Sun has flipped its polar magnetic fields to the opposite polarity consistently 5 years into each solar cycle. This has never failed until now.

The Sun's polar magnetic fields should have flipped in 2013 for solar cycle 24, but nothing happened in the northern region. The polarity flipped in the southern region in 2015, but in the North, the polarity from the previous cycle fizzed out to nothing. The north of the Sun became magnetically neutral. It remained that way.

We should see a big blue patch starting in 2013 in the North, as in cycle 22. But nothing happened? How weak has the solar system become for the polar magnetic reversal to fail?

As the Sun is understood in mainstream science

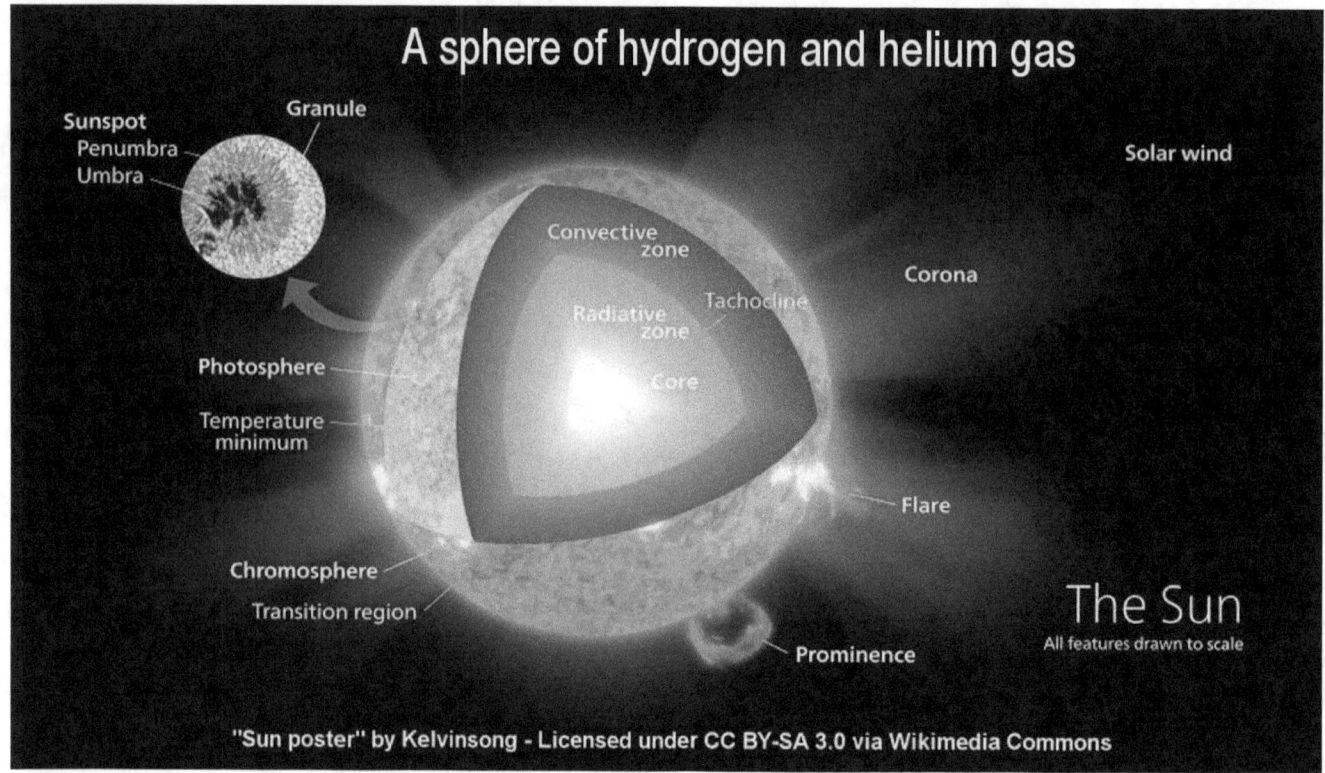

Obviously, this amazing event isn't caused by the Sun itself. As the Sun is understood in mainstream science, it cannot suddenly change so dramatically.

Putin had something to do with this

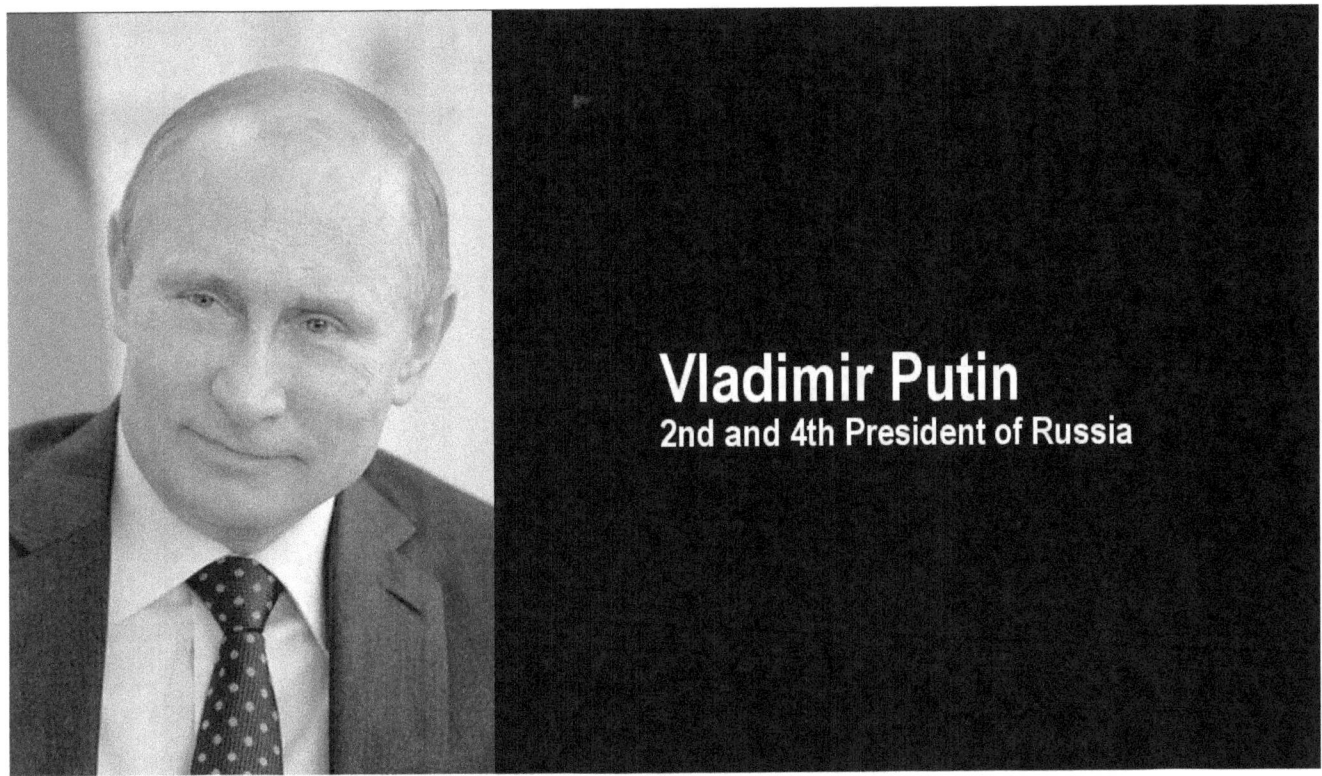

The only logical explanation that remains, therefore, is that the Russian President Vladimir Putin had something to do with this, as we say in the West when anything bad happens in the world.

We could also ask the Sun

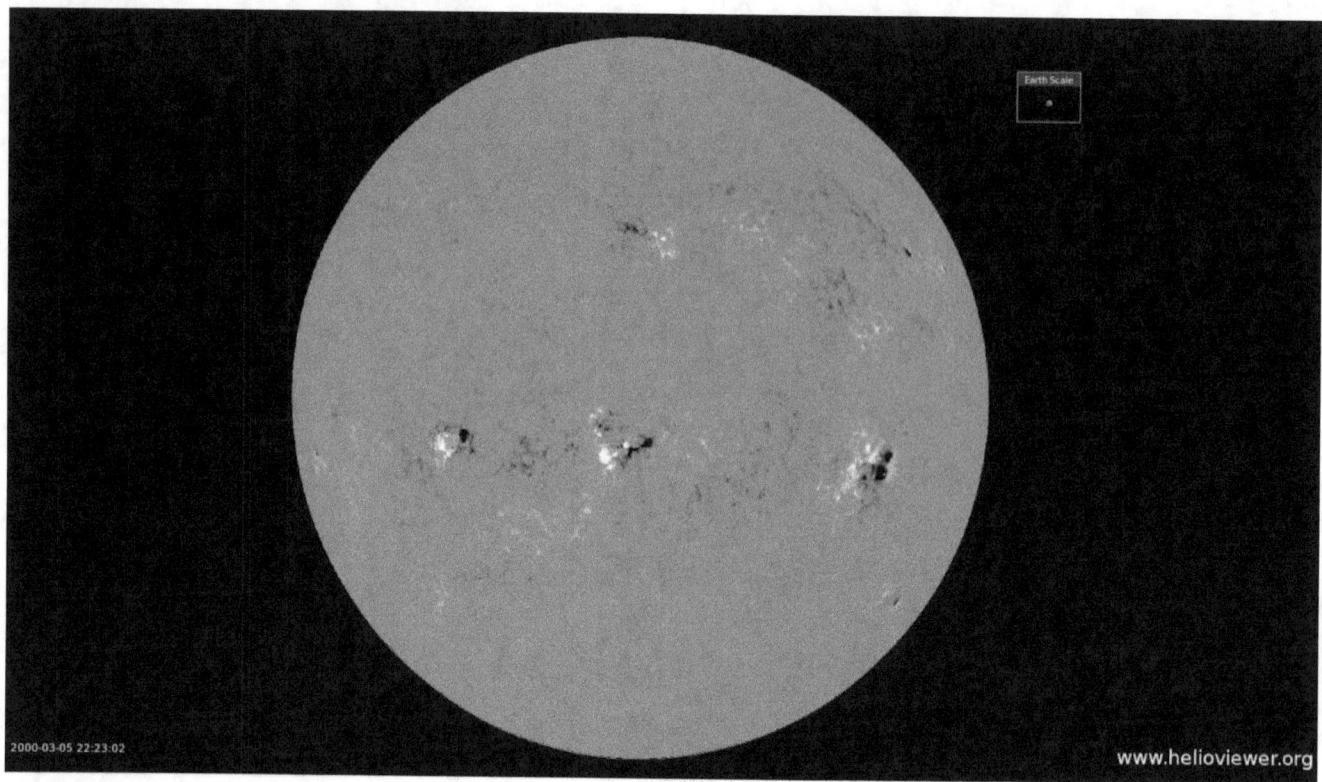

Of course, we could also ask the Sun, about what IT can tell us about the incidence. It might shed some light on the case. All we need to do is decipher its language.

A 19-page research paper on the subject

The reversal of the Sun's magnetic field in cycle 24

Institute of Solar-Terrestrial Physics,
Russian Academy of Sciences, Irkutsk, Russian Federation.
 - Alexander V. Mordvinov

National Solar Observatory, Sunspot, New Mexico 88349, USA.
 - Alexei A. Pevtsov

National Solar Observatory, Tucson, Arizona, USA.
 - Luca Bertello,
 - Gordon J.D. Petrie

https://arxiv.org/ftp/arxiv/papers/1602/1602.02460.pdf

Analysis of synoptic data from the Vector Stokes Magnetograph (VSM) of the Synoptic Optical Long-term Investigations of the Sun (SOLIS) and the NASA/NSO Spectromagnetograph (SPM) at the NSO/Kitt Peak Vacuum Telescope facility

We know that what happened is real. When a number of top scientists from three leading institutions, with access to data from the most advanced technological instruments, became so concerned that they issued a 19-page research paper on the subject, it should raise some eyebrows, and probably more than that.

To look under its skin

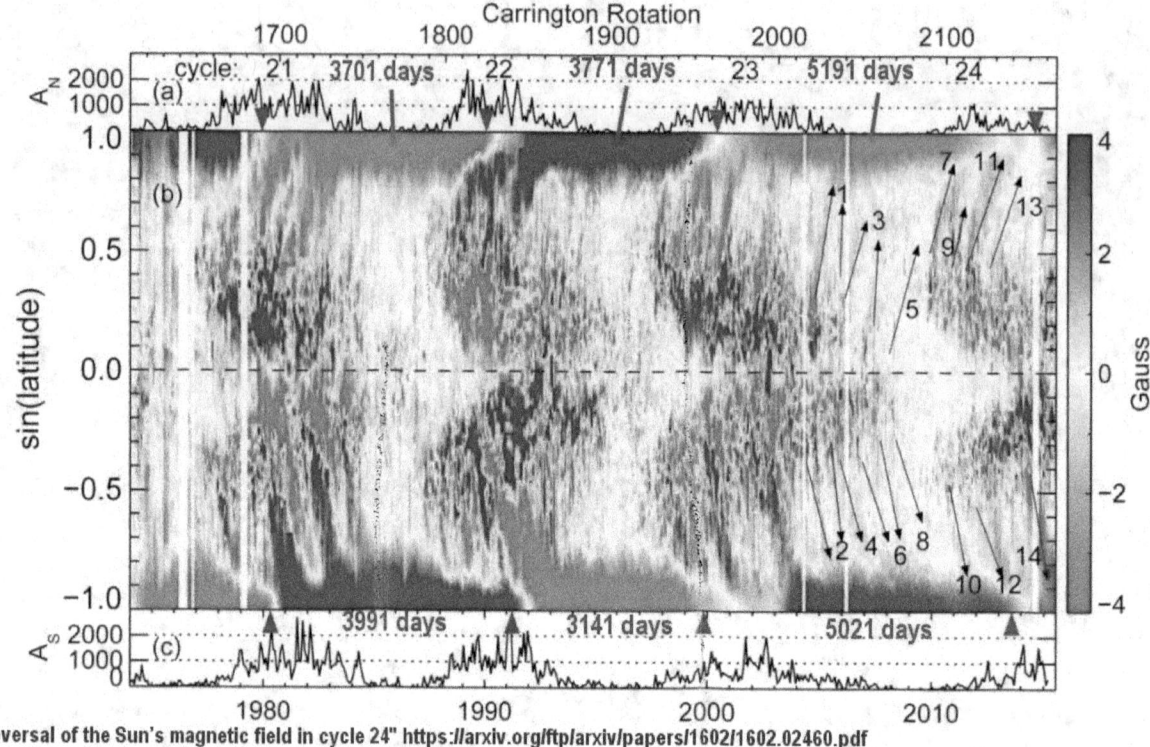

"The reversal of the Sun's magnetic field in cycle 24" https://arxiv.org/ftp/arxiv/papers/1602/1602.02460.pdf

The theories that are presented in the research report are largely speculative for the lack of a clear understanding of the language of the Sun, in spite of the amazing technologies that we now have for looking at the Sun.

The data that we now have should enable us to look deeper than we have in the past, to look under its skin, and to look beyond even that, to the principles involved for what we see, as I am doing with this video.

Can we learn the language of the Sun?

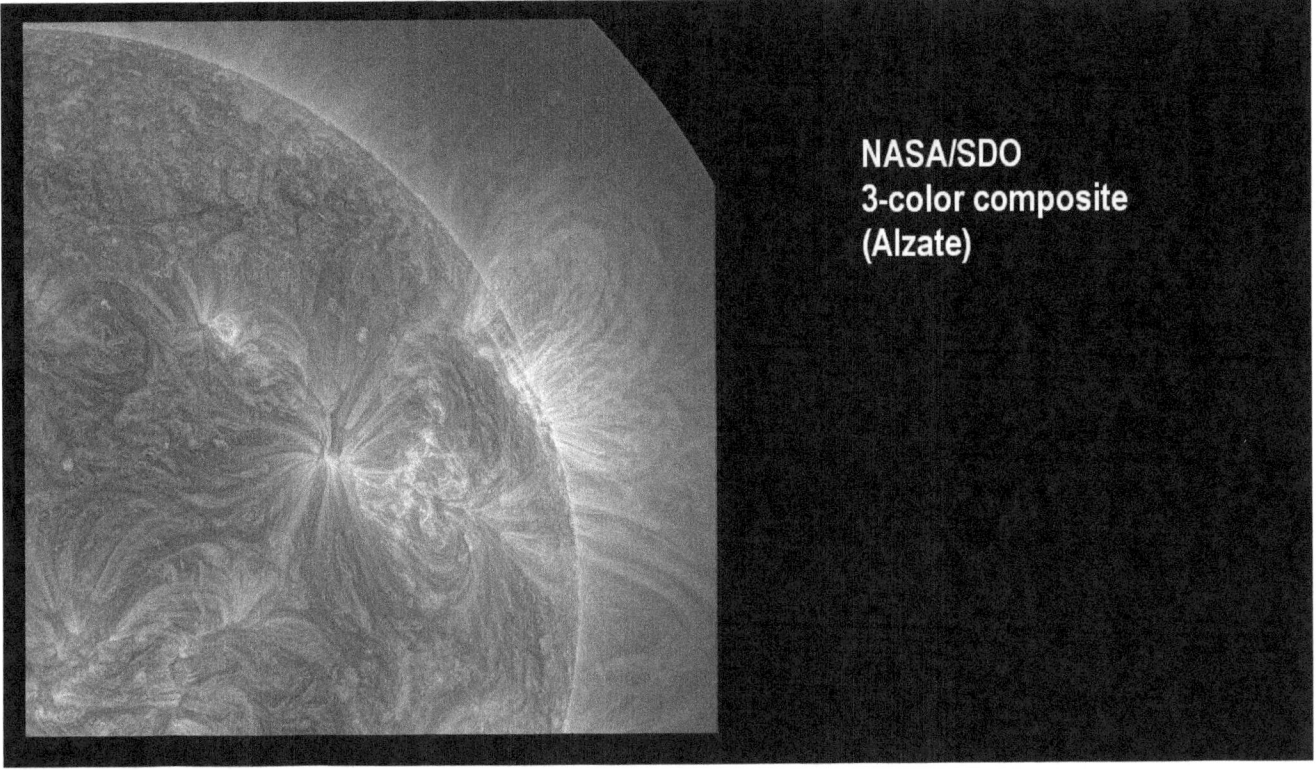

But is this really possible? Can we learn the language of the Sun? Can we decipher what the Sun is really telling us with all its numerous hints and paradoxes, from the gigantic to the minuscule? Maybe we ought to make a greater effort than ever before, to look really under its skin and learn its language from the ground up, to gleam its secrets that have so far puzzled us.

Human civilization became developed into what it is today, on this platform, with the development of languages. Shouldn't this experience help us with the Sun, too, so that we may be able to discover the cosmic technologies that make it tick?

Advanced deciphering of the 'language' of the Sun

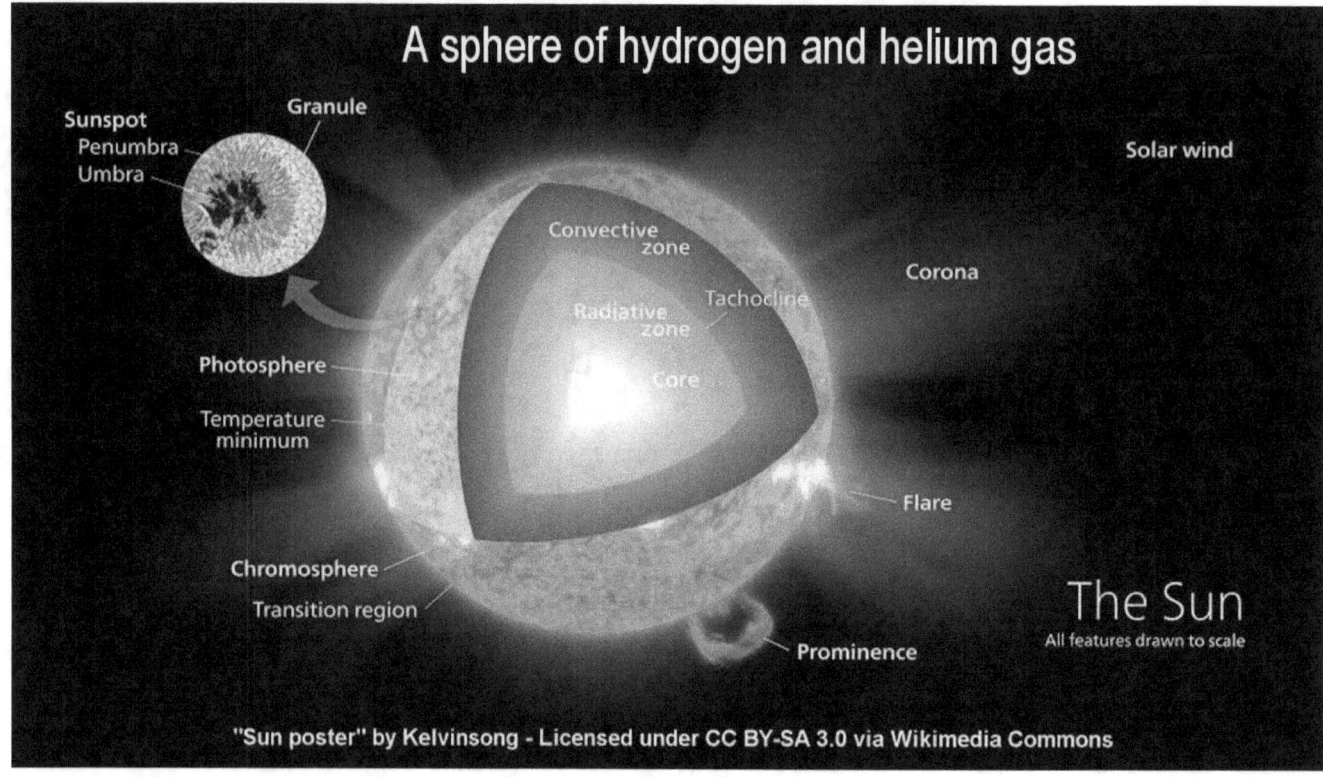

Shouldn't an advanced deciphering of the 'language' of the Sun help us at the very least to step away from the far-too-many mythological concepts that we still harbour about the Sun?

Ignoring what the Sun has to tell us

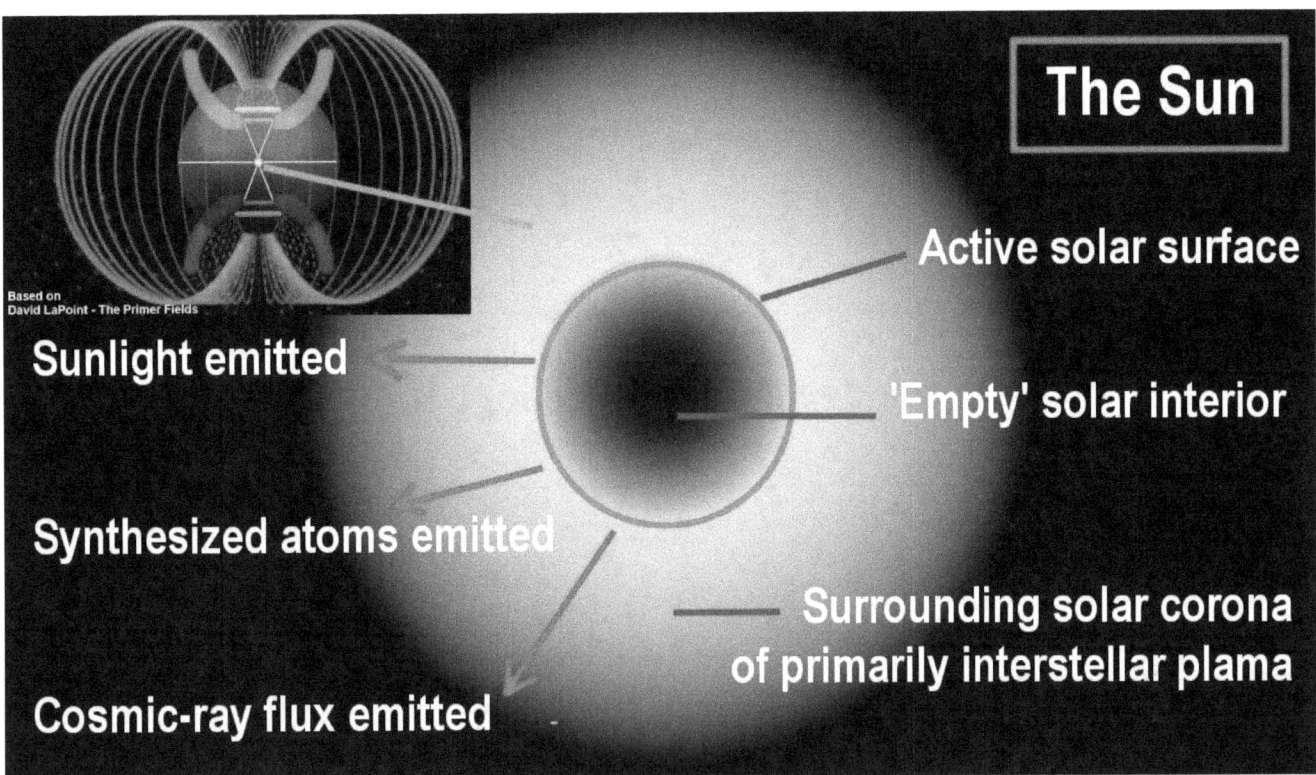

This would be a critical step, because as we will discover thereby, we are presently locked into a war against our very existence by ignoring what the Sun actually has to tell us about it.

➢ On the development of languages

On the development of languages

On the development of languages

Development of written languages

The development of culture and civilization goes hand in hand with the development of written languages and their power of expression, to communicate ideas and discovered principles.

Some languages have become lost

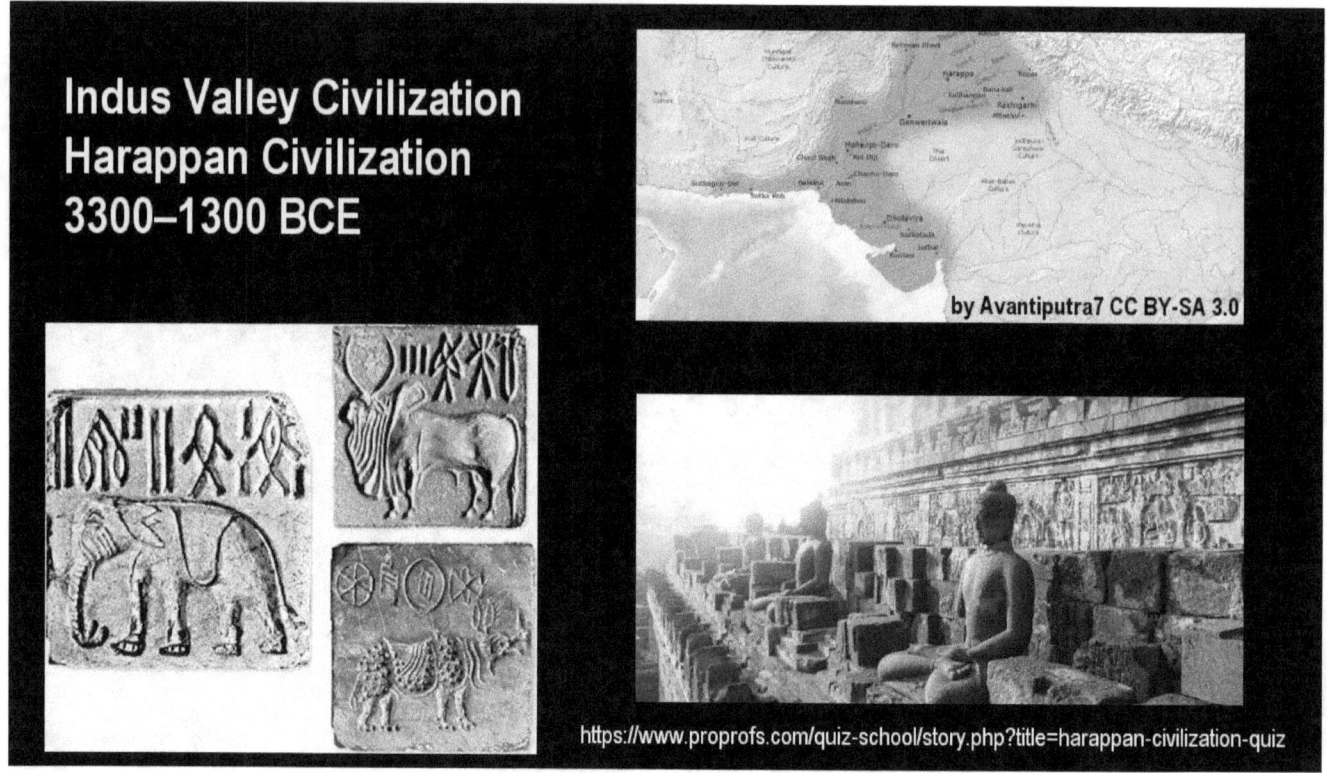

Some languages had flourished and the societies with them. And some have become lost over time so that we can no longer decode them.

An example of the lost language is the language of the Harappan Civilization that had flourished amazingly well in the Indus Valley for 2,000 years. Its language, though it became a written language, became lost when the civilization suddenly vanished 3,300 years ago.

Epic literary works attributed to Homer

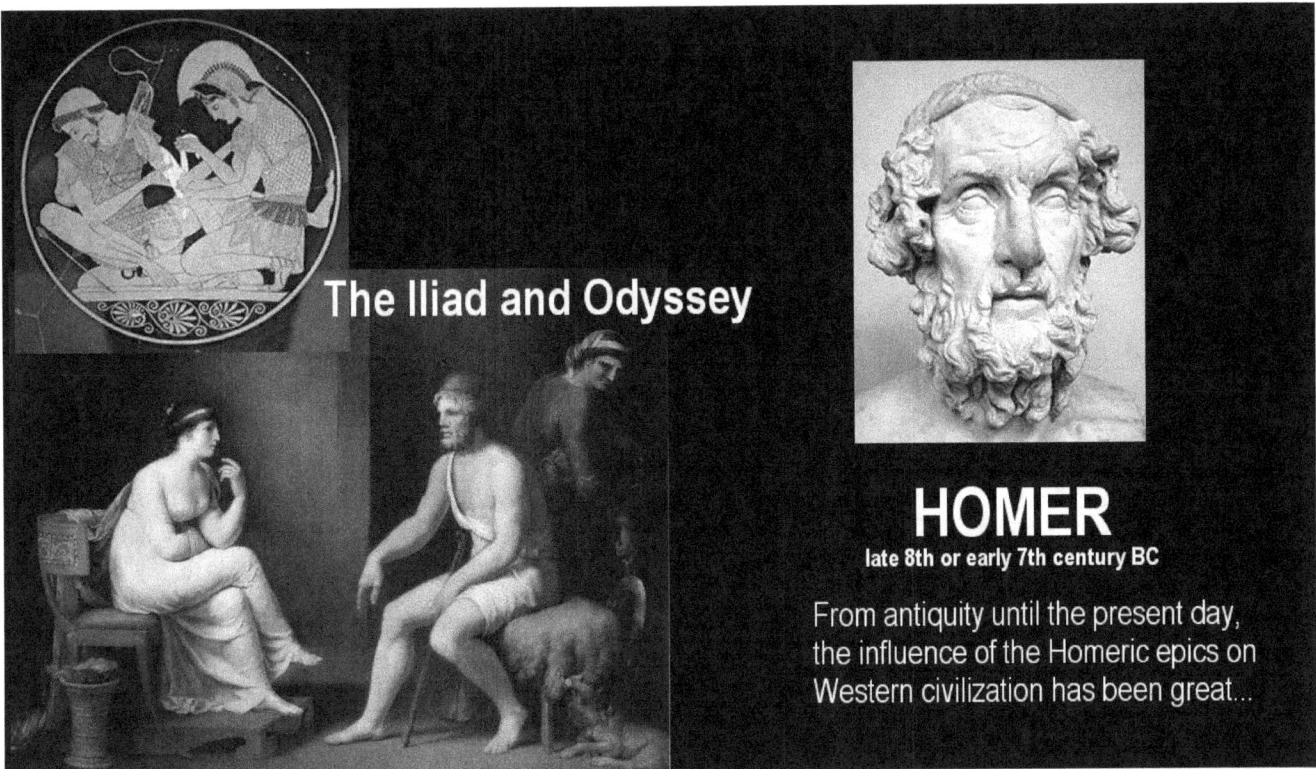

It appears that all great cultural advances in the history of civilization had an equally great written linguistic developments standing behind it, such as the epic literary works attributed to Homer, which has set the stage for the dawn of the Greek Classical culture, and has shaped Western Culture up to modern time.

Language development by Dante Alighieri

Illustration for Paradiso (of The Divine Comedy) by Paul Gustave Louis Christophe Doré

Dante Alighieri, detail from Luca Signorelli's fresco, Chapel of San Brizio,

It may well be that without the language development efforts by Dante Alighieri, back in the 12th Century, who brought together the best features of the Italian dialects into a comprehensive high-level language, the Golden Renaissance in the 15th Century might not have happened, and the great humanist developments that came out of it might not have occurred.

Cosmic universe has a specific 'language'

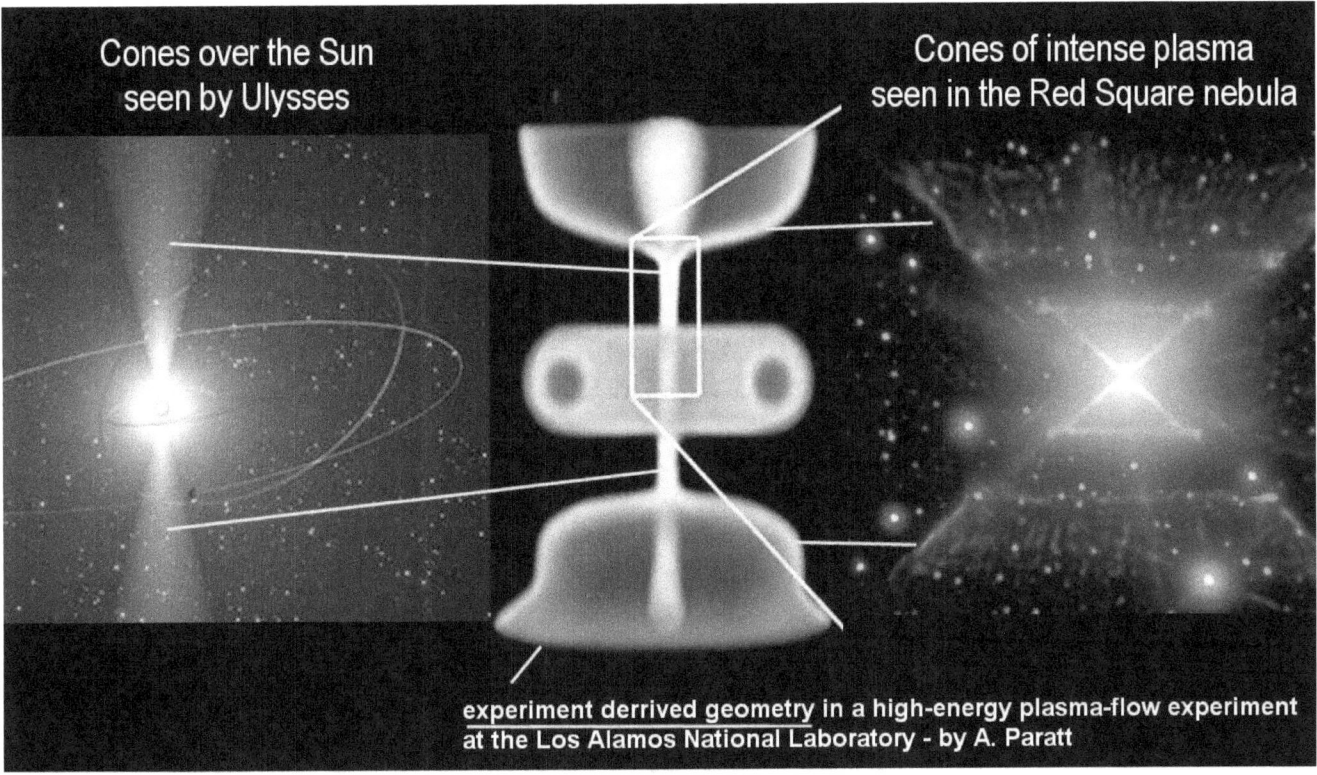

It appears that even the cosmic universe has a specific 'language' as it were, through which it can be understood.

➢ **Universe 'speaks' the 'language' of plasma**

The Universe 'speaks' the 'language' of plasma

a 'language' that modern society is gradually beginning to decipher

The Universe 'speaks' the 'language' of plasma

a 'language' that modern society is gradually beginning to decipher.

At the Los Alamos National Laboratory

Researchers at the Los Alamos National Laboratory have come to recognize that more than 99.999% of the mass of the universe exists in the form of plasma. They have come to recognize that plasma spans all space; and is the lifeblood of the universe, hence its name, plasma.

Plasma, simply is everywhere

Plasma, simply is everywhere. It is so small that it cannot be seen, but is so basic to everything that even the portion of the universe that isn't plasma, is a construct of it.

Plasma is a part of the universe in its most intimate explicate form, as the leading theoretical physicist, David Bohm, sees the universe. He sees the apparently empty cosmic space in terms of a vast sea of latent energy, with ripples in its expression unfolding that become discrete, but not separate from it.

David Bohm, as his successor

Albert Einstein is quoted, to have referred to David Bohm, as his successor.

Called protons and electrons

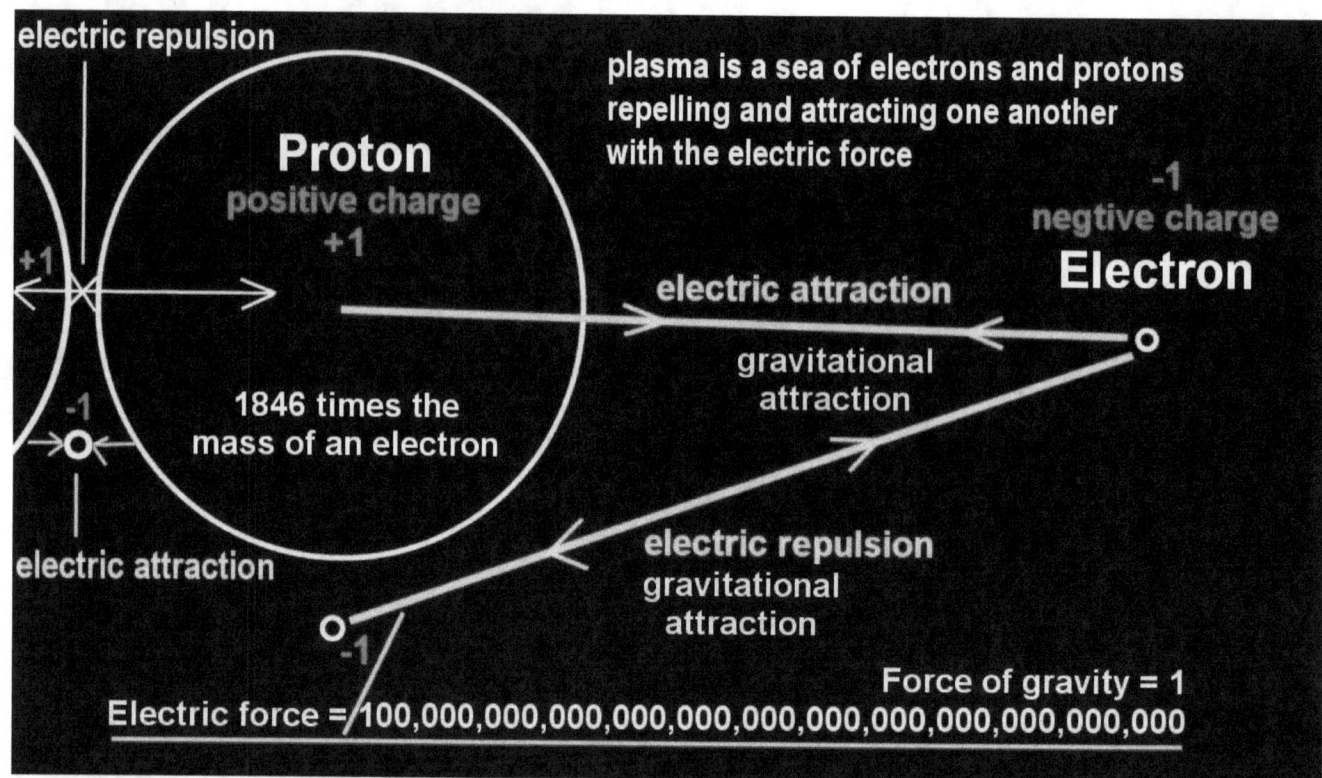

Plasma exists as a mixture of two types of tiny, electrically charged particles, called protons and electrons, that interact with the immensely powerful electric force that is 39 orders of magnitude stronger than gravity.

Electrons are the smallest components

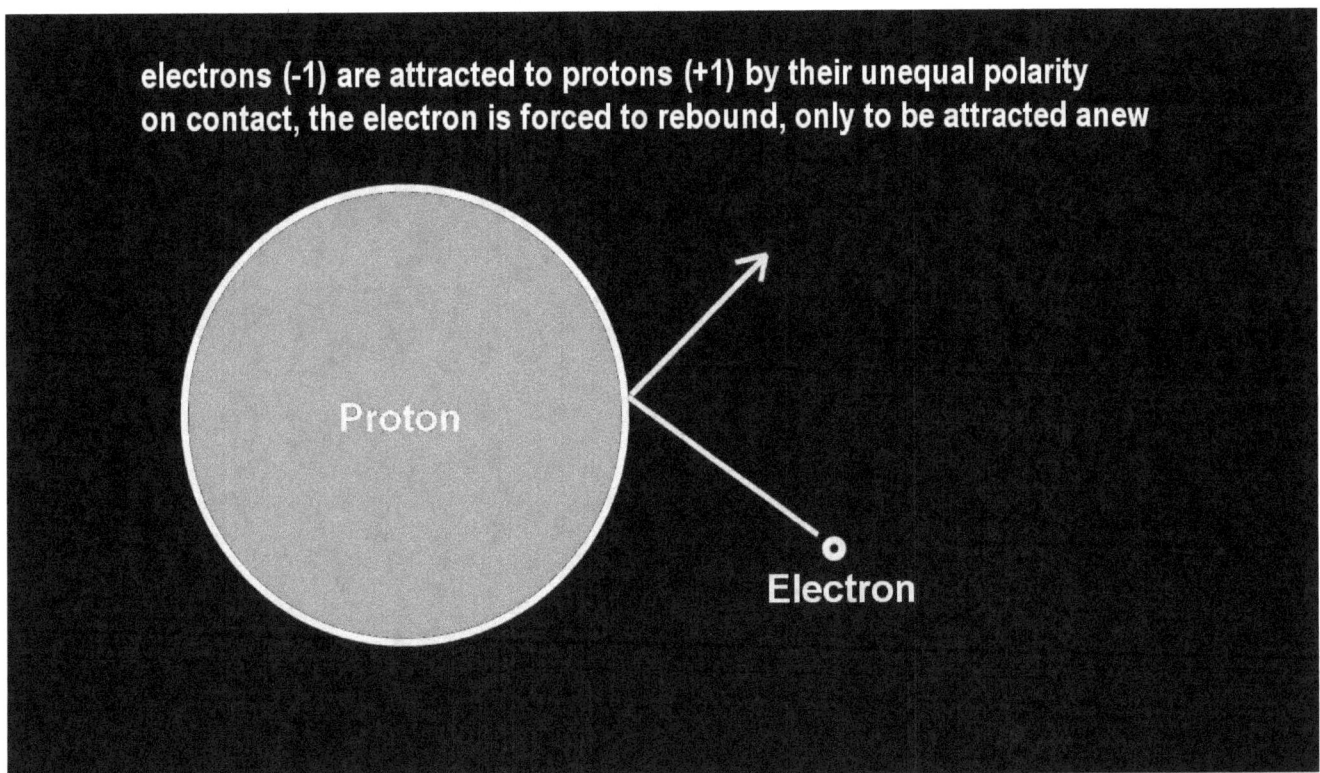

The electrons are the smallest components of plasma. They are almost 2,000 times smaller than the protons. They are electrically attracted to the protons, but then are strongly repelled on contact by an even stronger nuclear force, only to be attracted again. Thus, the electrons are drawn into an endless dance around and between the protons.

Cosmic space is filled with examples

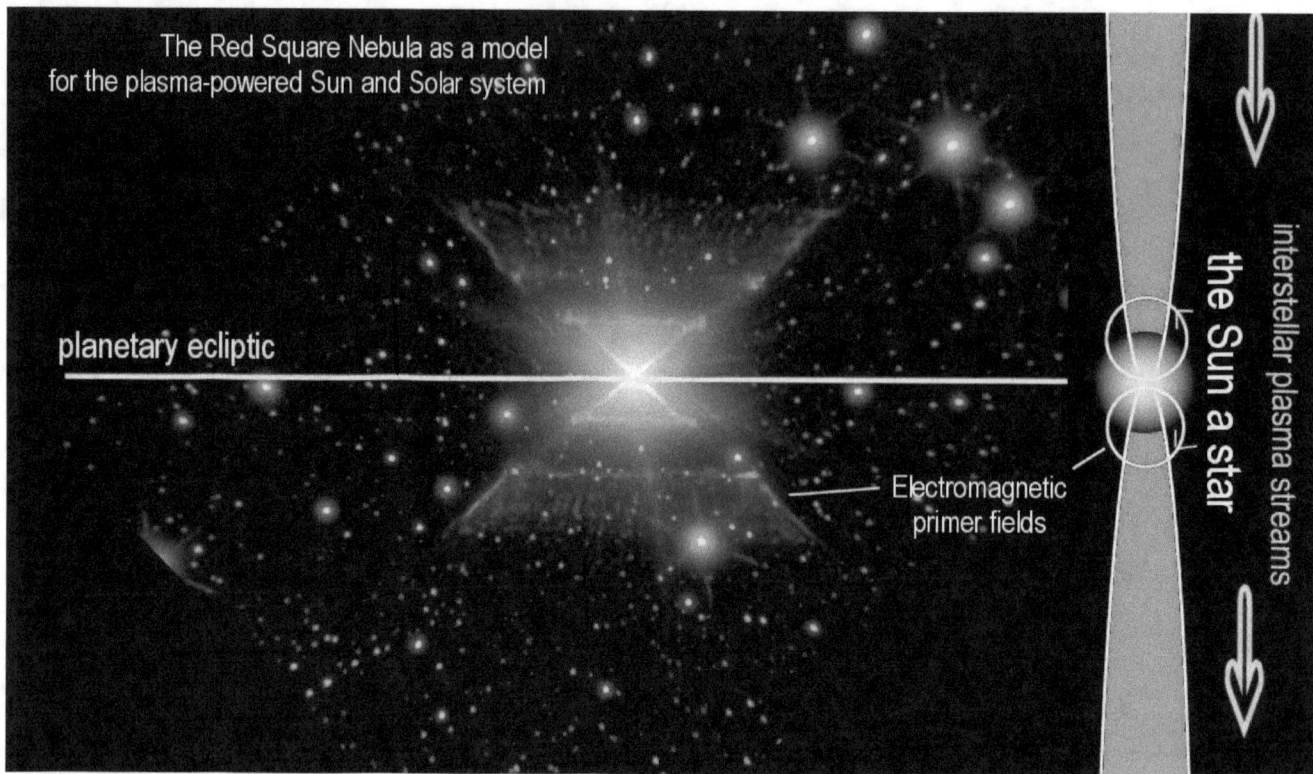

By the interaction of free-flowing plasma in cosmic space, plasma combines into moving streams. In some cases plasma also combines into large coherent spheres that interact with the moving streams. Cosmic space is filled with examples of both.

➢ The plasma stream

> The plasma stream

The plasma stream

It speaks the language of magnetism

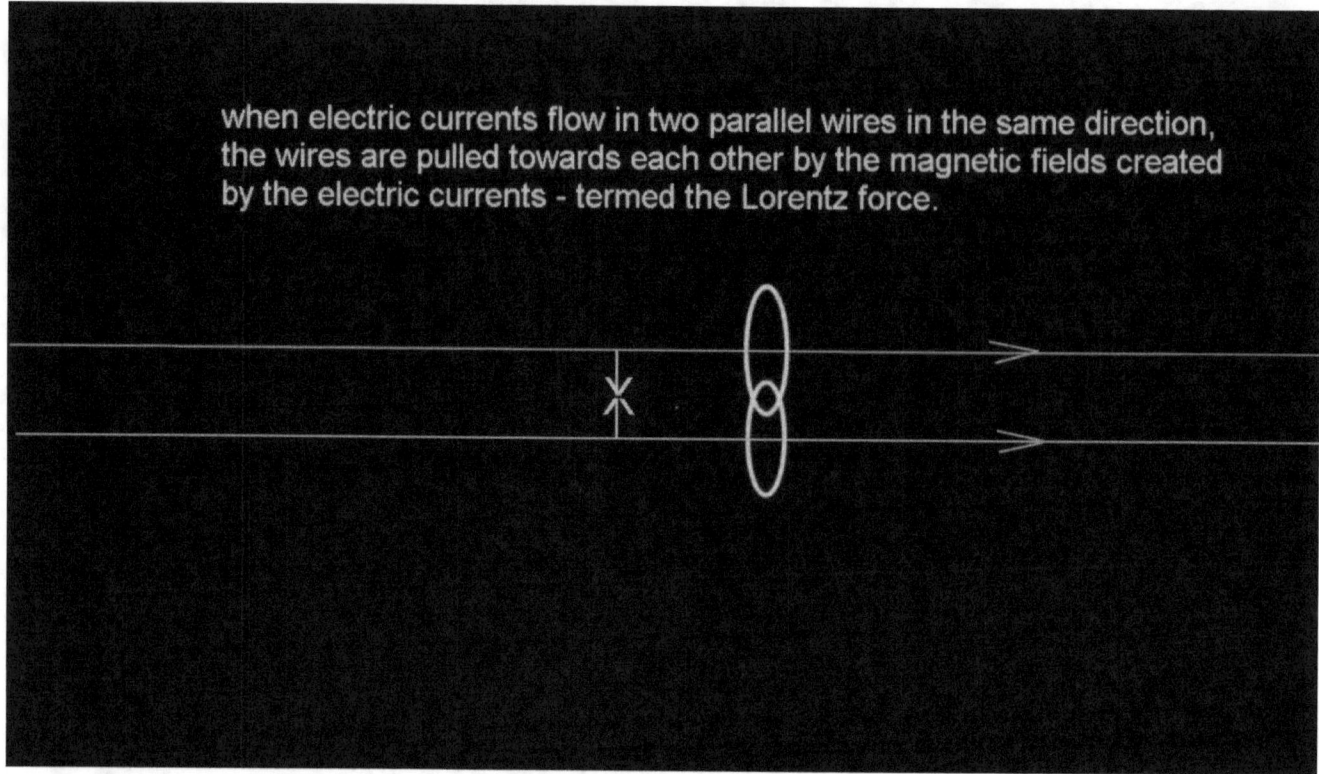

When plasma is in motion, it speaks the language of magnetism. The movement of electrically charged particles creates a magnetic force around the path of the motion.

When electrons flow in two parallel wires in the same direction, the encircling magnetic fields combine and draw the wires towards each other.

Termed the Lorenz force

The effect is termed the Lorenz force, in honour of its discoverer.

Wires in cosmic space

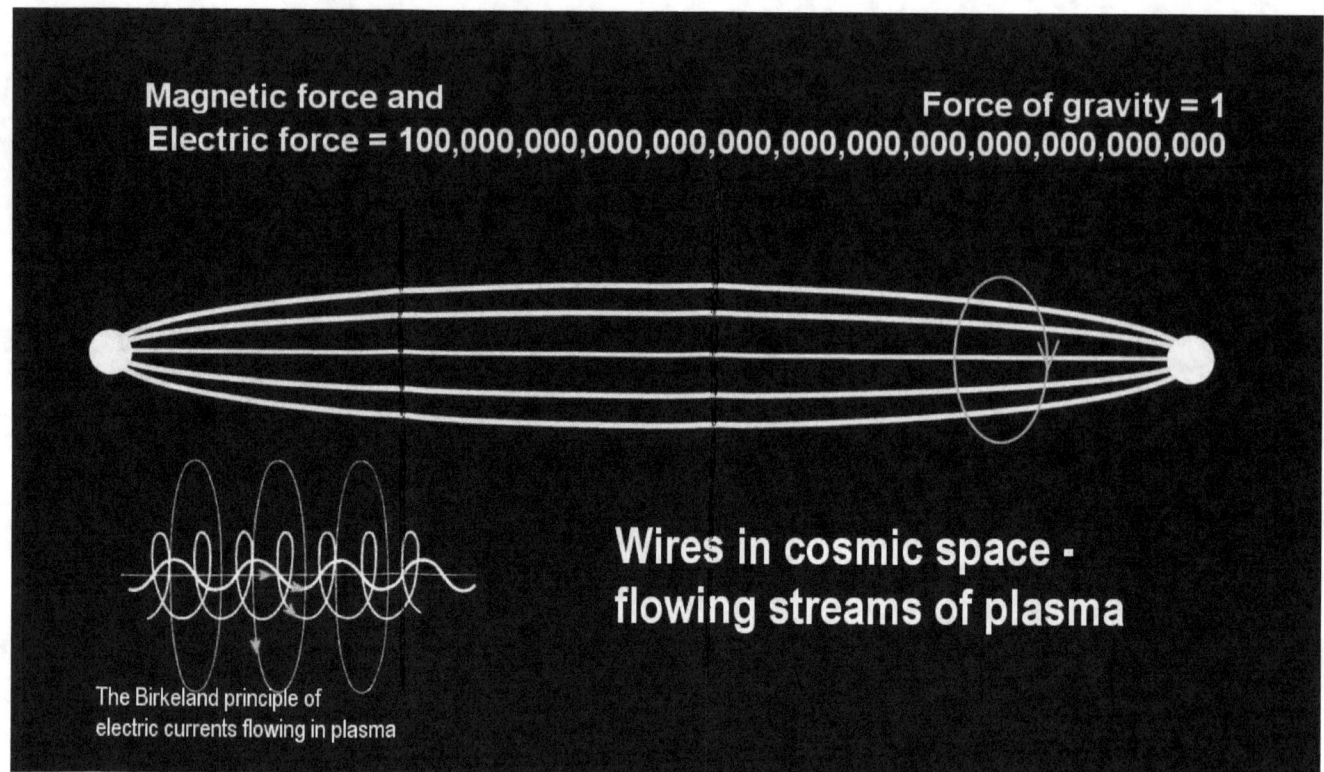

Wires in cosmic space -

flowing streams of plasma

In space, the combining effect of the electric and magnetic fields, draws large movements of plasma together into self-concentrating streams. These self-aligned streams can extend across large distances, spanning light years between stars, even millions of light years between galaxies, and clusters of galaxies, and so on.

As the plasma streams flow

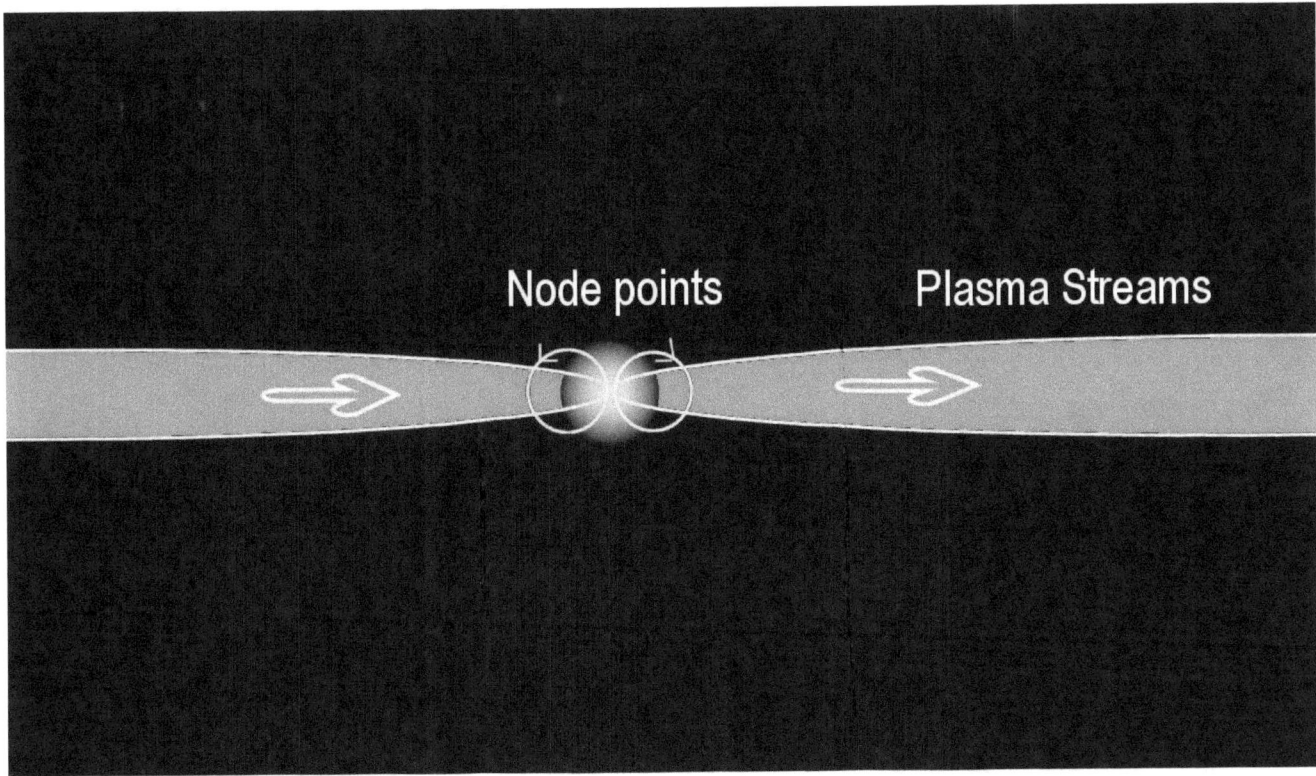

Also, as the plasma streams flow, they attract plasma from surrounding space, whereby they increase in density. And, as they gain ever-greater density, the plasma streams become pinched together more strongly, magnetically. Of course, as the cross section of the stream becomes smaller by the pinch effect, the resulting greater compression density also increases the magnetic field concentration, which in turn increases the pinch effect.

The self-escalating process of pinching the plasma into ever-smaller volumes, continues until the interacting system creates itself a crisis and breaks down. At this point a different set of principles come into play, according to the unfolding situation.

The amazing plasma-flow geometry

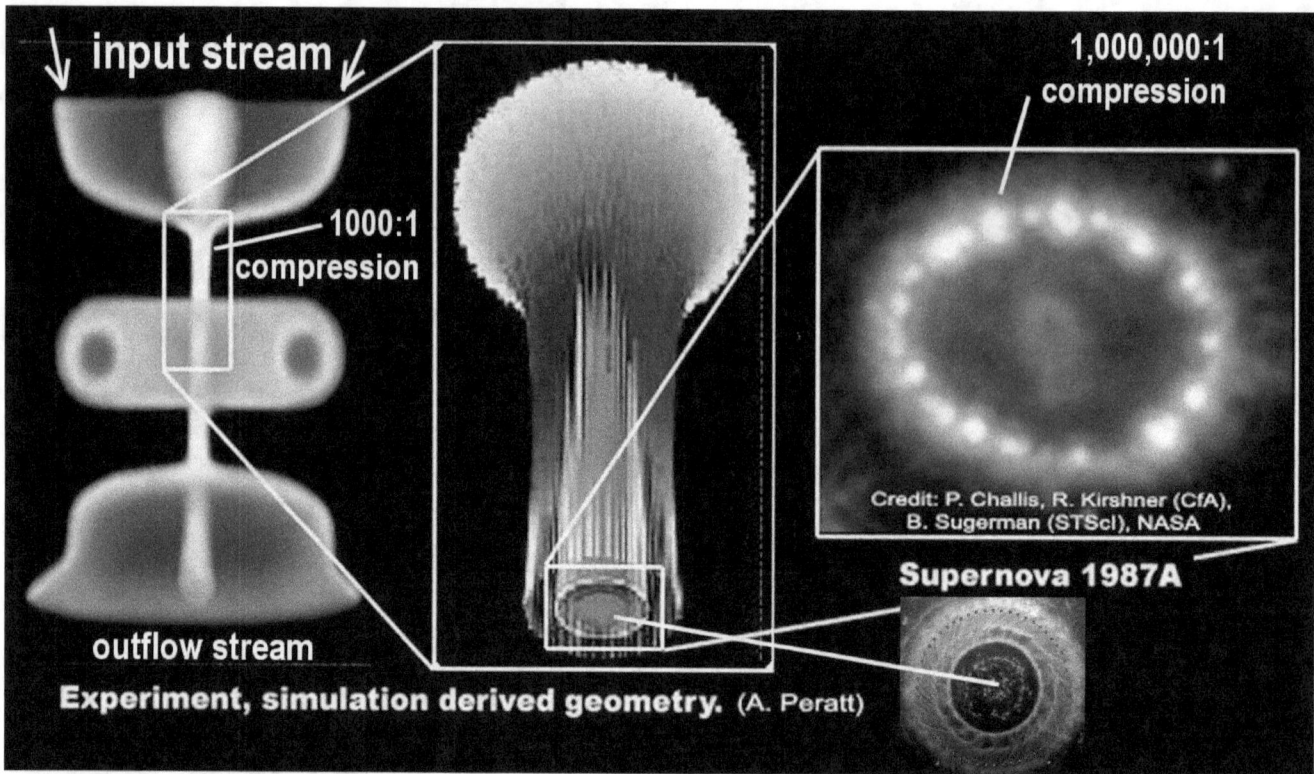

In order to discover what happens at the point of crisis, a set of high-energy discharge experiments were conducted at the Los Alamos National Laboratory, in New Mexico in the USA, under the direction of Anthony Peratt.

The amazing plasma-flow geometry that is illustrated here, shown in purple, resulted in the experiment. The entire plasma input stream was concentrated into an extremely narrow stream of great density. A compression ratio exceeding 1000:1 resulted in the experiment, possibly as high as 5000:1.

The experiments further revealed that the narrow stream that resulted, was itself but a ring-like construction of 56 discrete filaments of plasma of extremely high density. The compression ratio at this point might have exceeded the range of a million to one.

An the illustration of a ring of filaments of this type, is found in cosmic space surrounding the center of the long-vanished Supernova 1987a.

The experiment illustrates, that when the pinch effect becomes so great that the plasma stream looses its forward momentum, it begins to flow into itself, whereby a new magnetic geometry unfolds that curls some of the moving plasma backwards, and which also folds the highly concentrated magnetic field backwards, by which the plasma becomes trapped under a dome-type structure, where it is super-concentrated, as it has no place to go.

Eventually, when the concentration becomes strong enough to overcome the confining force under the dome, a narrow stream of hyper-concentrated plasma flows explosively out of the confinement.

He termed, the Primer Fields

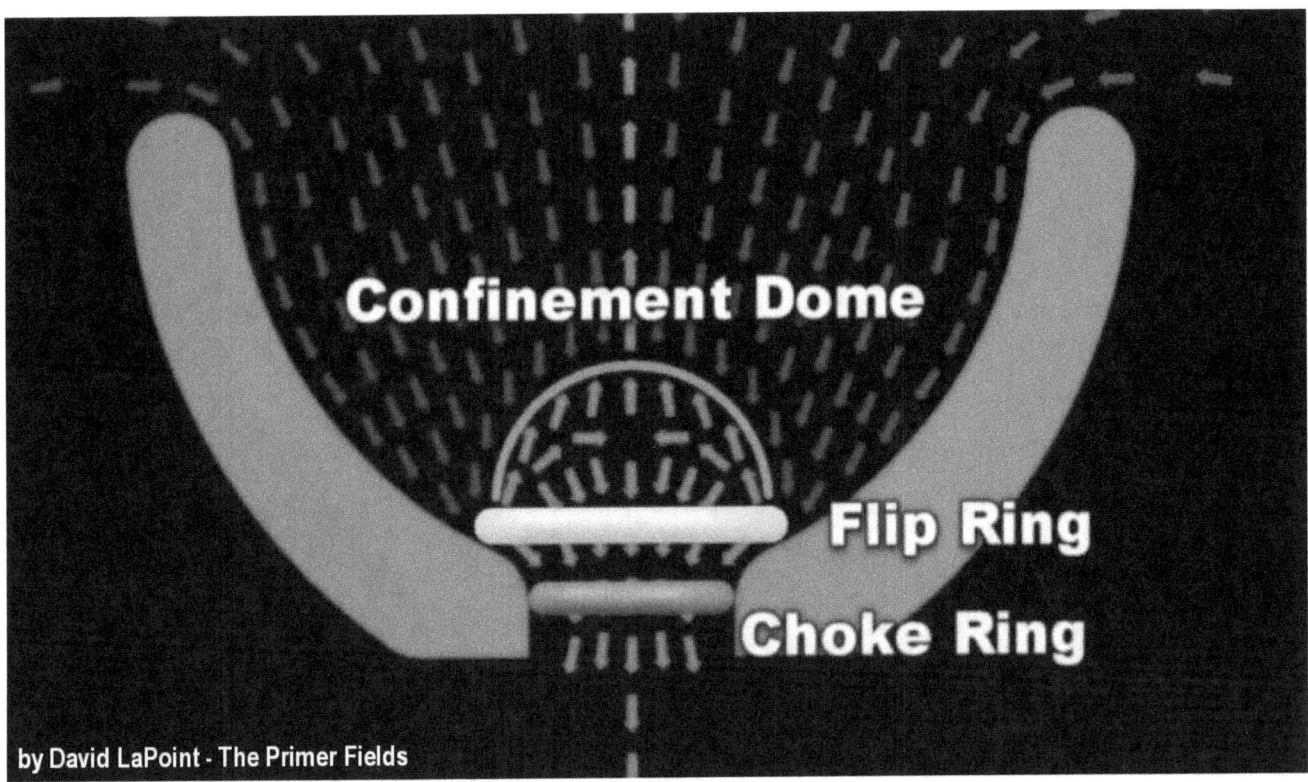

The plasma-flow researcher, David LaPoint, who explored the dynamics statically, termed the magnetic confinement structure, the Confinement Dome, and the magnetic field that flips the plasma backwards, the Flip Ring, and another magnetic field that keeps the plasma from escaping, the Choke Ring. And the entire complex, he termed, the Primer Fields.

Enormously large magnetic forces

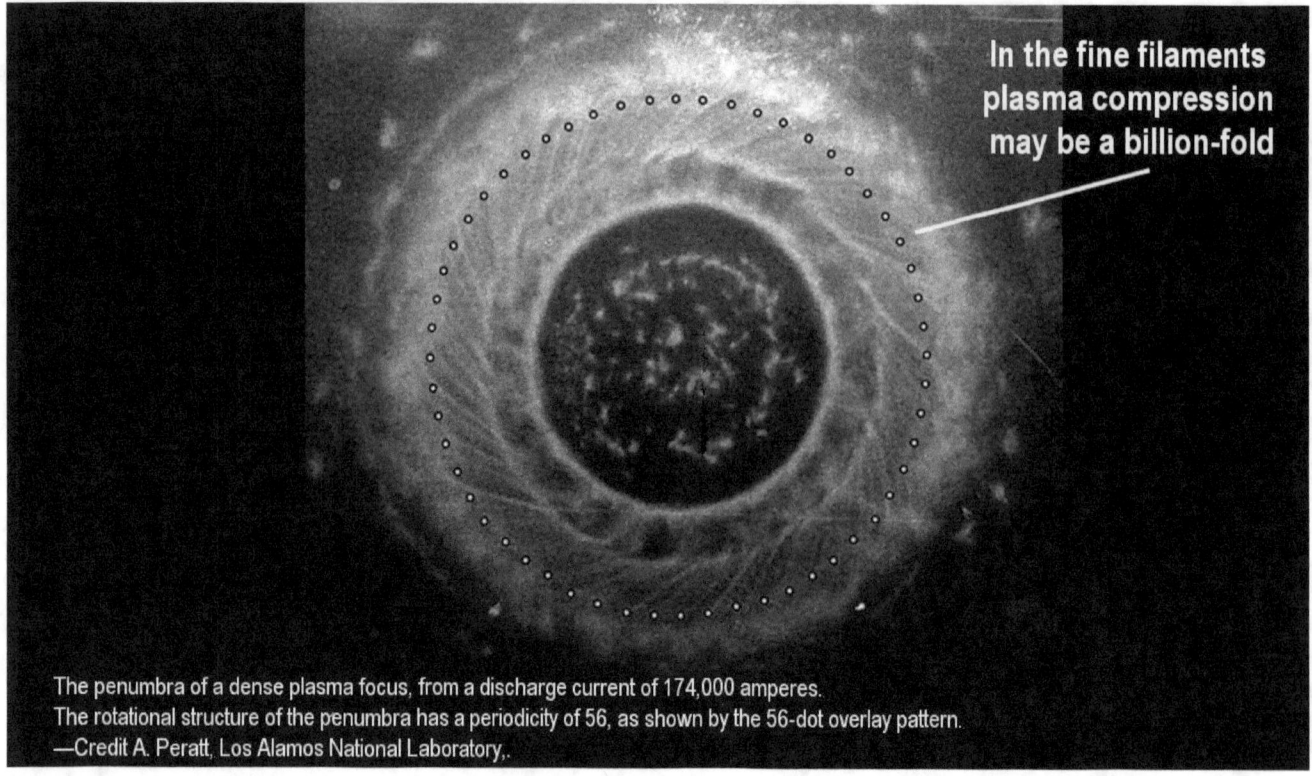

In the fine filaments plasma compression may be a billion-fold

The penumbra of a dense plasma focus, from a discharge current of 174,000 amperes.
The rotational structure of the penumbra has a periodicity of 56, as shown by the 56-dot overlay pattern.
—Credit A. Peratt, Los Alamos National Laboratory,.

Enormously large magnetic forces must have developed in the process to create the filaments and to keep the entire construct as tightly confined as the experiment illustrates.

➤ The plasma sphere

> The plasma sphere

The plasma sphere

Plasma is not visible

Plasma is not visible. It doesn't emit light, nor does it absorb it. Plasma becomes visible only by its effect, such as by moving plasma energizing atomic elements, that thereby emit light, and create sunlight.

When large volumes of plasma form a sphere

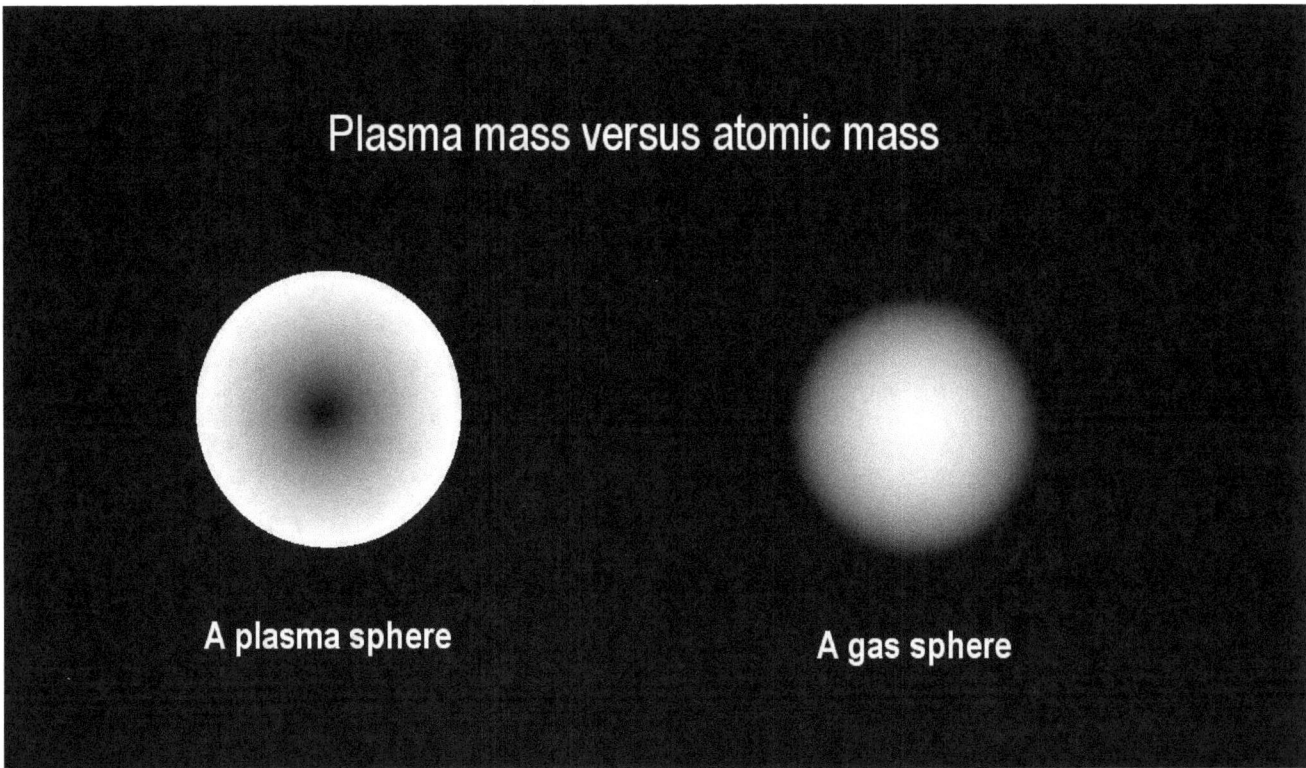

When large volumes of plasma form a sphere, combined by gravity, the electrons of the plasma, which are in constant motion, being attracted to protons, but repelled again at close distances, and them being 2000 times lighter than the protons, tend to migrate to the surface, away from the center of gravity. As this happens, the protons that are more isolated thereby, repel each other more strongly. The end-result is, that a large sphere of plasma has its highest mass-density, and also the highest electron density, on its surface, and has the least of them at its center.

Such a sphere of plasma is a Sun

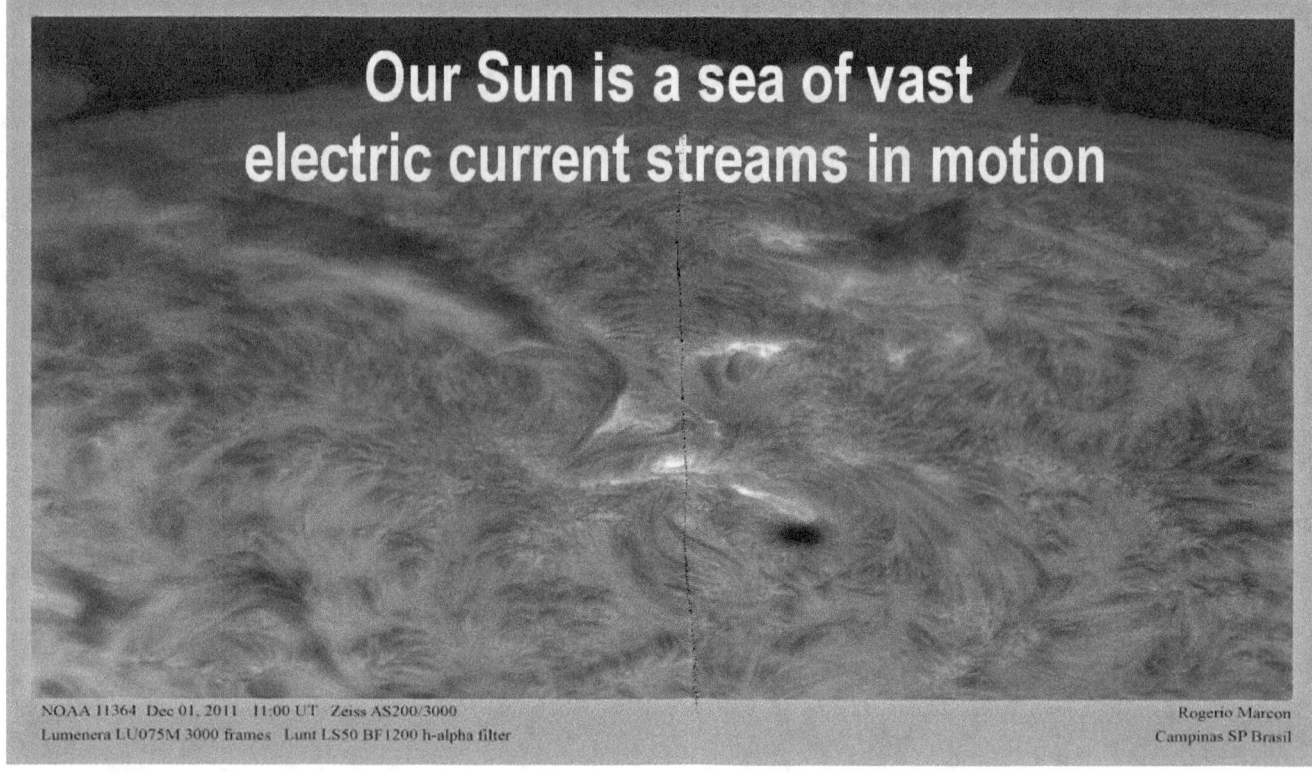

Such a sphere of plasma is a Sun. It exists as a relatively light shell of plasma, with little mass inside, but is teeming with a great density of electrons on its surface, forming a vast carpet of electric activity in the form of flowing streams of plasma.

This 'electrically charged' sphere

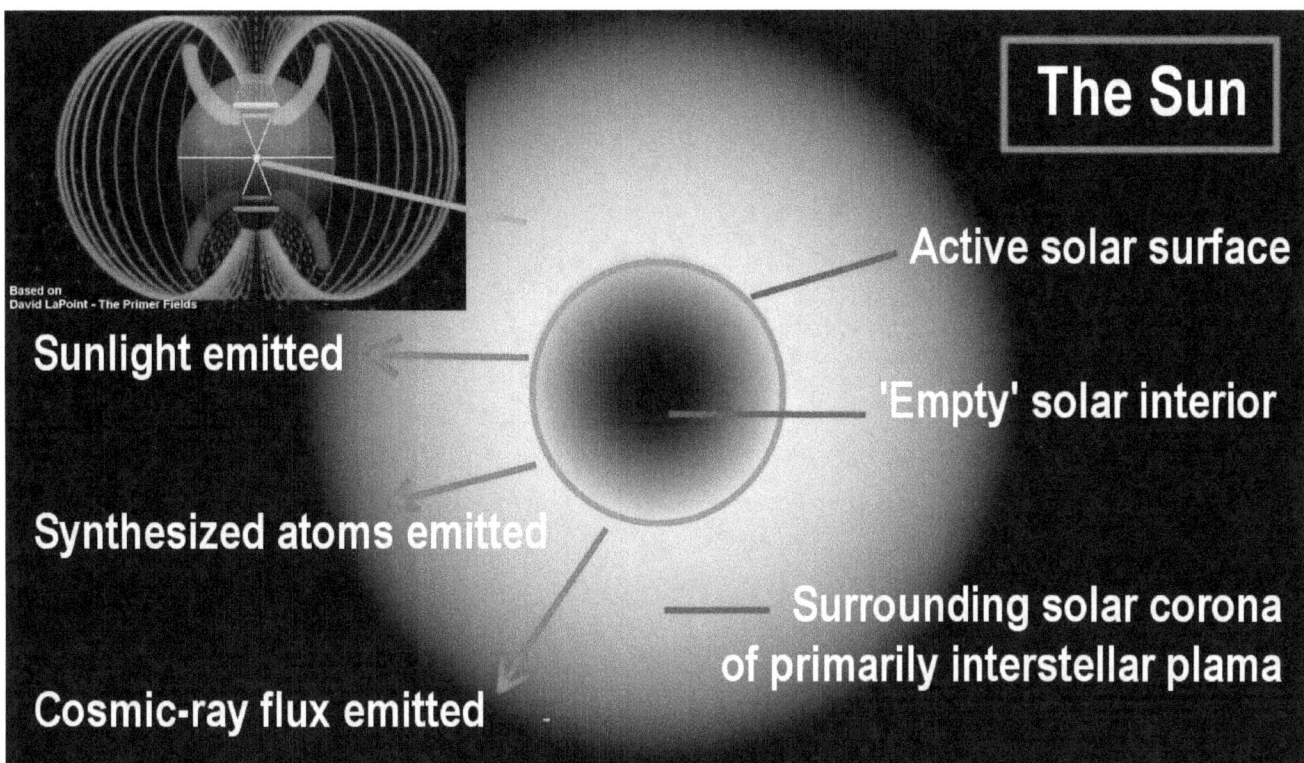

This 'electrically charged' sphere presents to the space around it an immensely strong electric field, with which it attracts plasma, and then reacts with it. It reacts with it so intensely, that atomic elements are synthesized in the process. In this immensely energetic process the synthesized atomic elements emit light. All atomic elements that exist are synthesized in this process, and emit light according to their individual spectra. The light is the sunlight that we see.

Every sun or star is a sphere of plasma. It reacts with interstellar plasma on its surface. Everything happens on the surface. Atomic elements are synthesized there and light is emitted by the atomic elements on a wide band of energy emissions that extend far above and below the range of the visible light band.

The Sun is dark inside

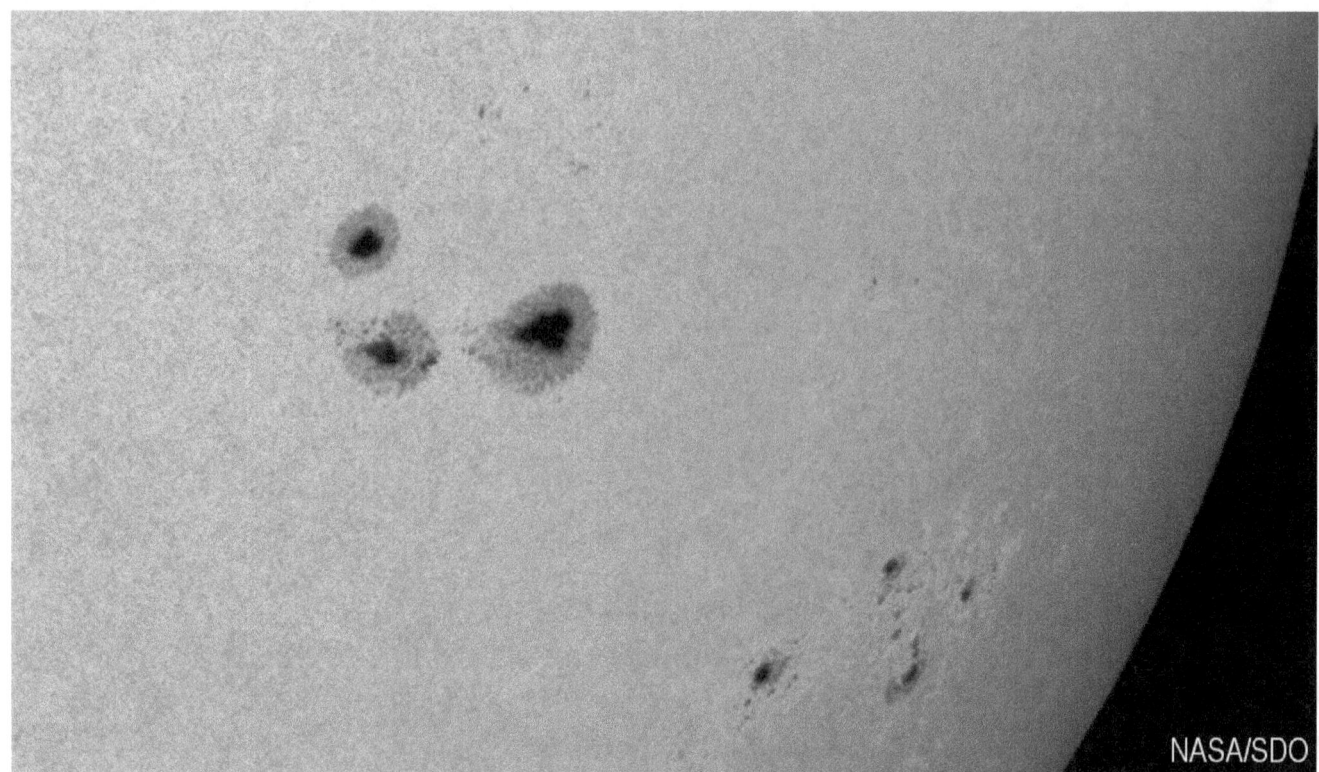

That this is so, is easily visible. Just look at the Sun with an appropriate telescope. The sunspots that you see are all dark at the umbra. The Sun is dark inside. Nothing happens within it. It is largely empty. But on the surface, the Sun has a granular structure that is a wide sea of plasma fusion cells.

Here is where the cosmic plasma stream phenomenon and the plasma sphere phenomenon come together.

The Sun comes to light as a star

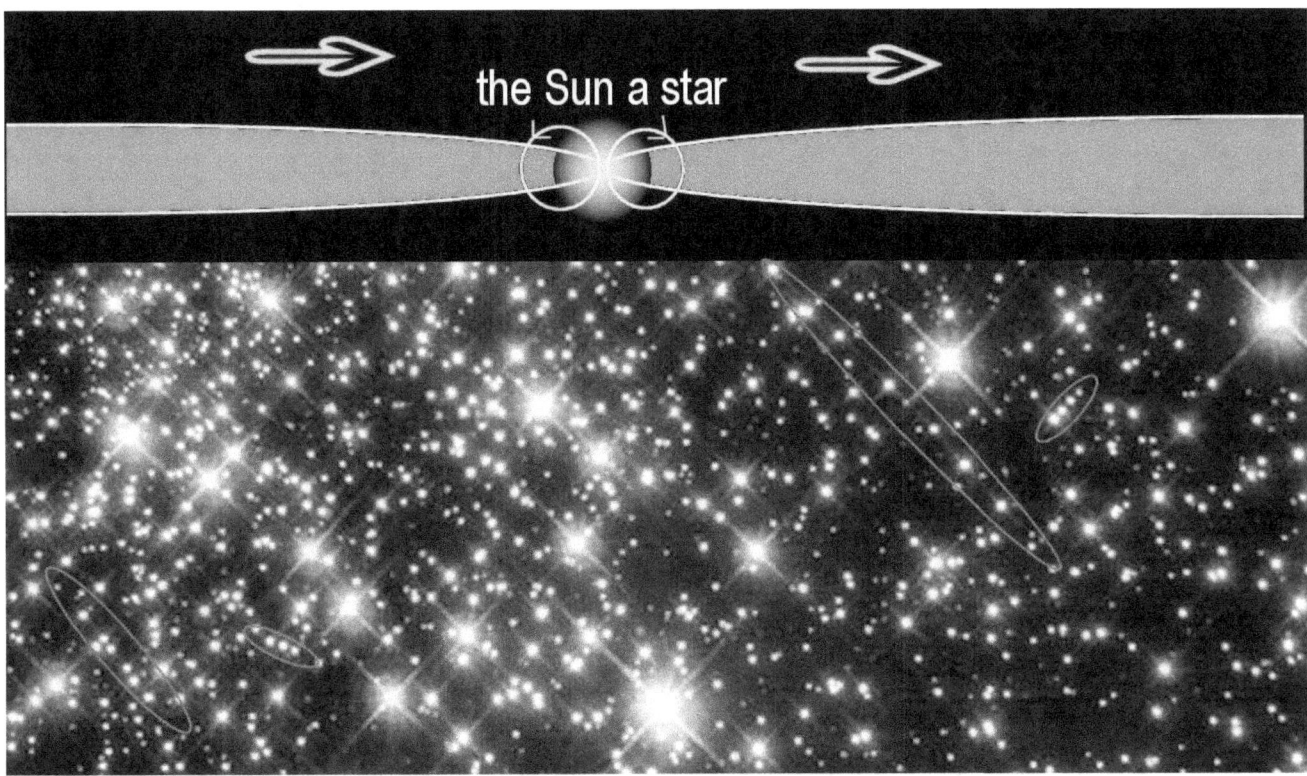

The Sun comes to light as a star that is located at a magnetic node point of an interstellar plasma stream, where the plasma in the stream becomes intensely concentrated.

The plasma Sun is located at the center

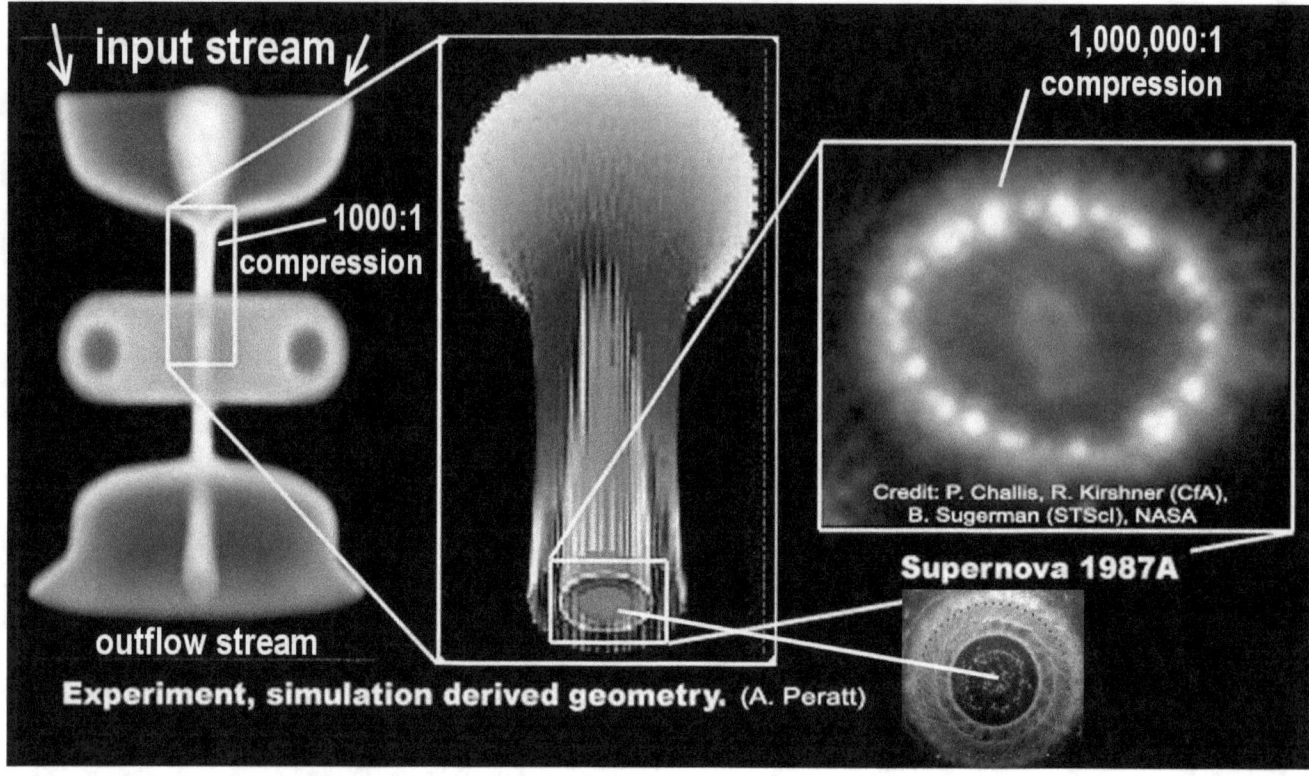

The plasma Sun is located at the center between two large magnetic primer field structures, similar to those created in the laboratory experiment.

In the experiment a portion of the plasma energy was dissipated into an ecliptic ring. The dissipation enables the plasma stream to expand again, which it does in the reverse of the process. It flows into a confinement dome, from which it expands again. Since the outflow is weaker, its confinement dome is correspondingly smaller.

Primer fields develop at a node point

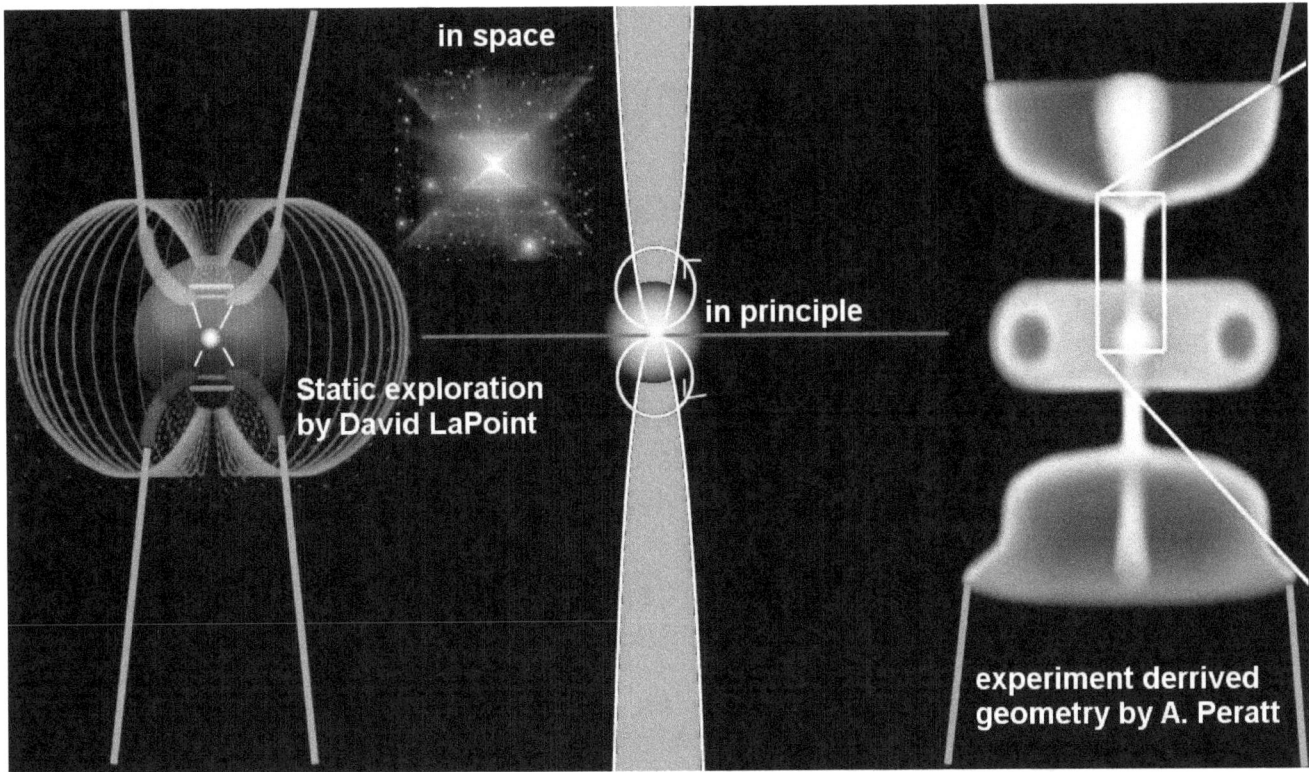

In the solar system, where these primer fields develop at a node point between two interstellar plasma streams, with the Sun in the middle, the Sun serves as the plasma-consuming element that makes the flow-through nature of the plasma structures possible. These flow-through type structures are not uncommon in space. They are typically seen in large nebula where enough atomic material exists in the background to make the plasma streams visible.

Let's decode the 'language' of the Sun

Let's decode the 'language' of the Sun

Unique Characteristics

> Different rotational speed of equator and poles
> Cyclical magnetic polarity reversals
> Symmetric magnetic polarity orientation
> Polar magnetic reversal delay
> Cyclical sunspot cycles
> Diminishing sunspot cycles
> Missing polar magnetic field

Let's decode the 'language' of the Sun

Unique Characteristics

Different rotational speed of equator and poles,

Cyclical magnetic polarity reversals,

Symmetric magnetic polarity orientation,

Polar magnetic reversal delay,

Cyclical sunspot cycles,

Diminishing sunspot cycles,

Missing polar magnetic field.

With the few basic concepts of the plasma 'language' established, we can begin to decode what the Sun is telling us about itself in the context of what we already know about it.

Sun rotates significantly faster at its equator

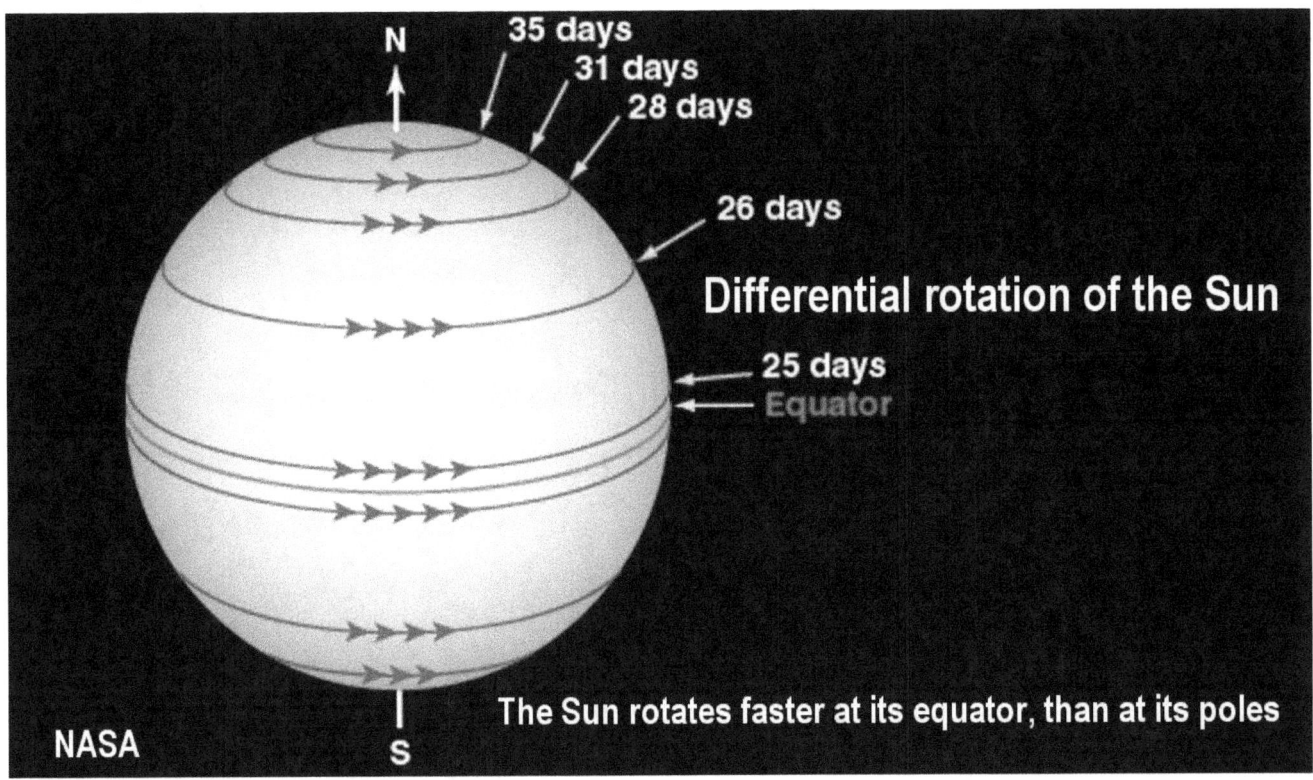

We know for example that the Sun rotates significantly faster at its equator than at its polar regions. The mechanistic Sun, as a gas sphere, doesn't allow for this to happen. But the plasma Sun tells us rather plainly that the differential rotation of the Sun is so natural that one would be surprised if it wasn't happening.

➢ The puzzling differential rotation

The puzzling differential rotation

The puzzling differential rotation

If the Sun was a body of atomic elements or gases

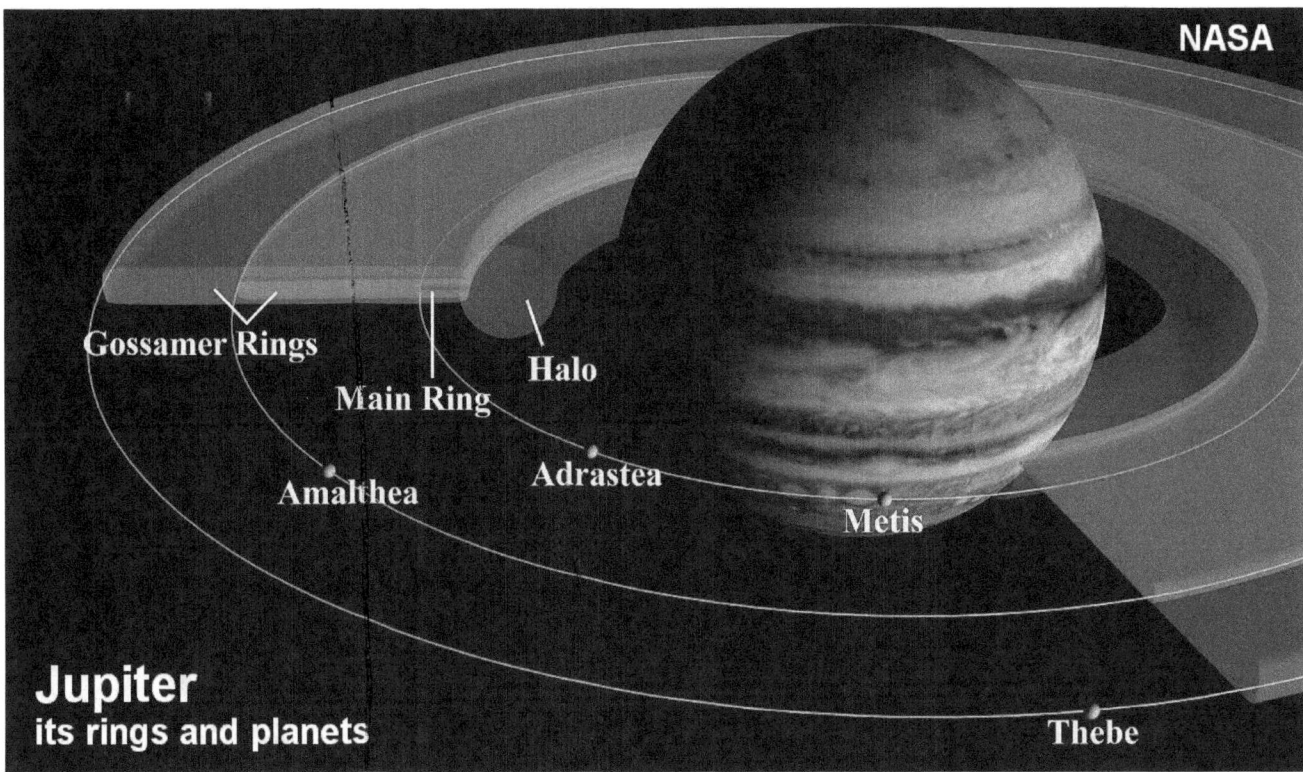

If the Sun was a body of atomic elements or gases, like the big gas planets are, it would, like them, rotate essentially as a single unit.

On Jupiter, the outer atmosphere rotates a minuscule 5 minutes slower at the poles, in its 10 hour rotation period. The difference adds up to 8/10th of a percent.

A 40% difference in rotational speed

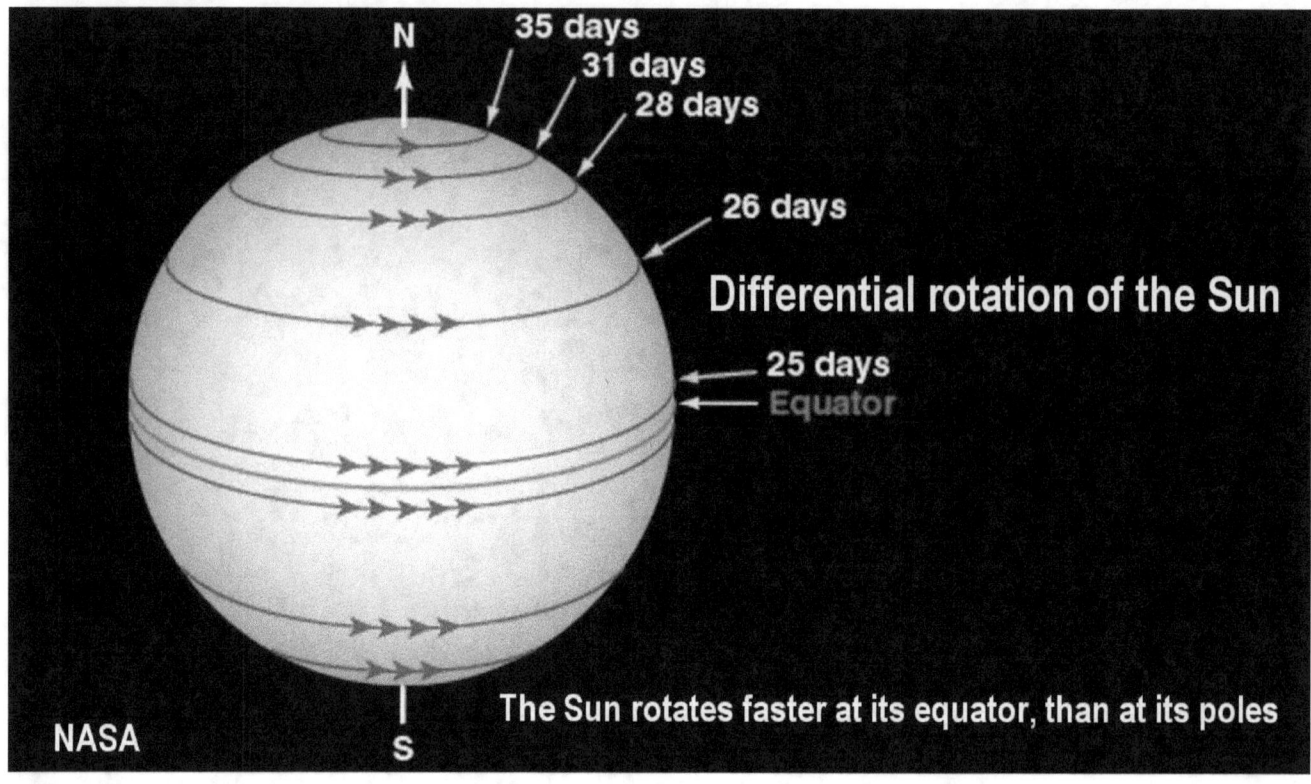

On the Sun, however, we see a 40% difference in rotational speed. The equator of the Sun rotates 40% faster. This difference is huge. And more amazing than this, is that the rotational speed varies only minutely across a wide band centered on the equator of the Sun, between the 30-degrees North and South latitudes. Across this band, the rotational speed varies only 4%, while the bulk of the difference in rotational speed occurs in the higher latitudes.

The large differential in the rotational speed of the Sun, is not readily visible. Nor is it of any great practical significance, except to indicate that the rotation of the Sun is externally powered by the rotating magnetic fields of the plasma stream in which it is located. No other cause, than an external cause, makes sense for this phenomenon; nor is an internal cause actually possible.

Rotation is driven from outside the Sun

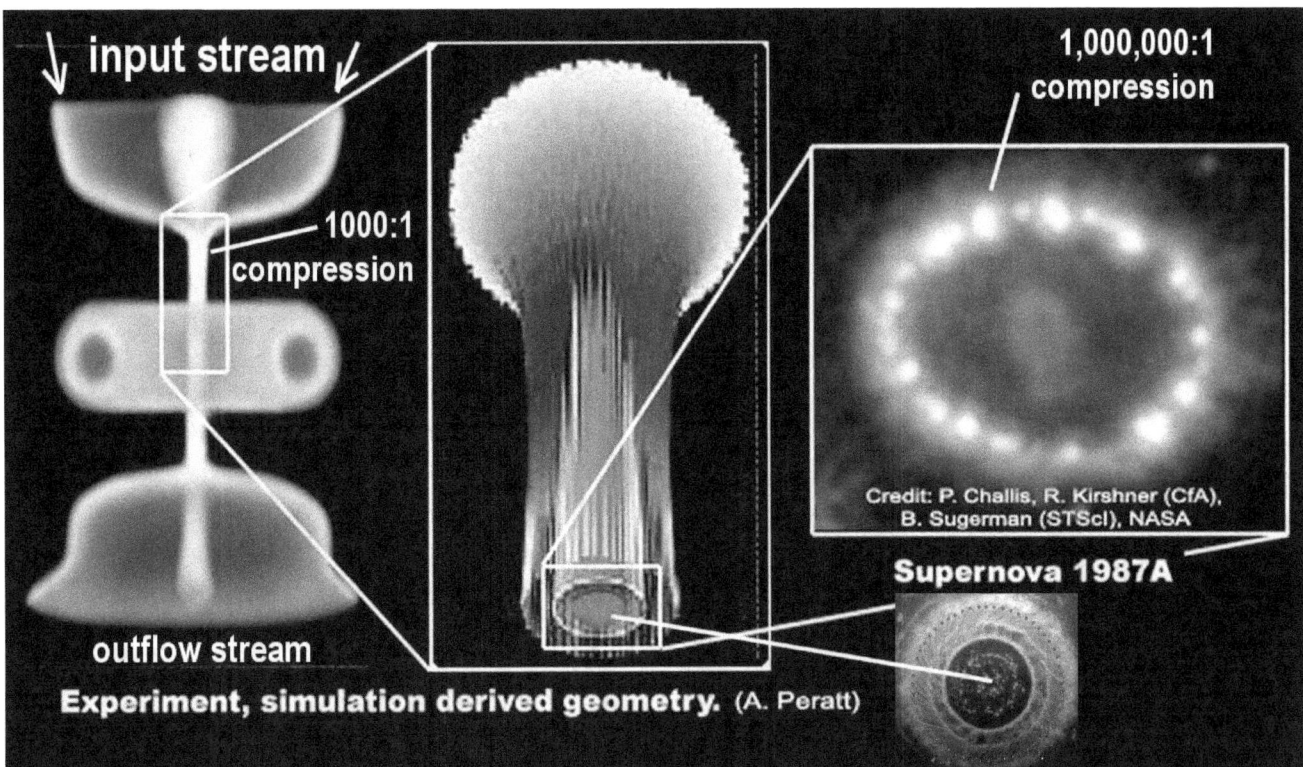

The answer why the differential rotation is driven from outside the Sun, can be found in the high-energy discharge experiment that I had previously referred to, that had been conducted at the Los Alamos National Laboratory by Anthony Peratt.

As I had noted earlier, the entire input stream became compressed in the experiment, by the self-formed primer fields, into a thin stream of plasma that incorporates a ring of 56 self-rotating plasma filaments. If one applies the observed principle to the solar system, and to the Sun being located within the high-density plasma stream, it becomes self-evident what powers the Sun's rotation.

The Sun located within a cylinder

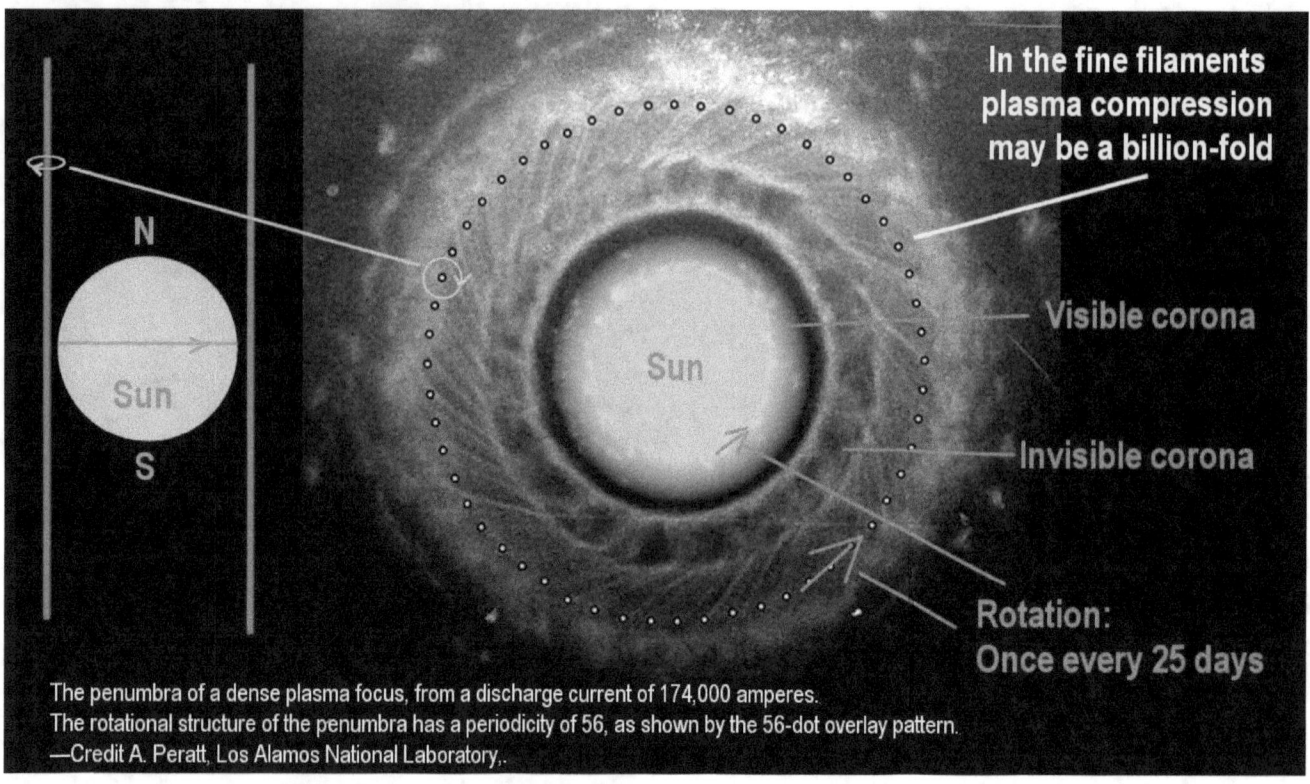

The penumbra of a dense plasma focus, from a discharge current of 174,000 amperes.
The rotational structure of the penumbra has a periodicity of 56, as shown by the 56-dot overlay pattern.
—Credit A. Peratt, Los Alamos National Laboratory,.

When one sees the Sun located within a cylinder of magnetically rotating plasma filaments, according to the principle of Birkeland currents, it becomes plain to recognize that the magnetic field that causes their rotation becomes inducted into the surface of the Sun.

Being plasma, the Sun is electrically conductive.

By electromagnetic coupling

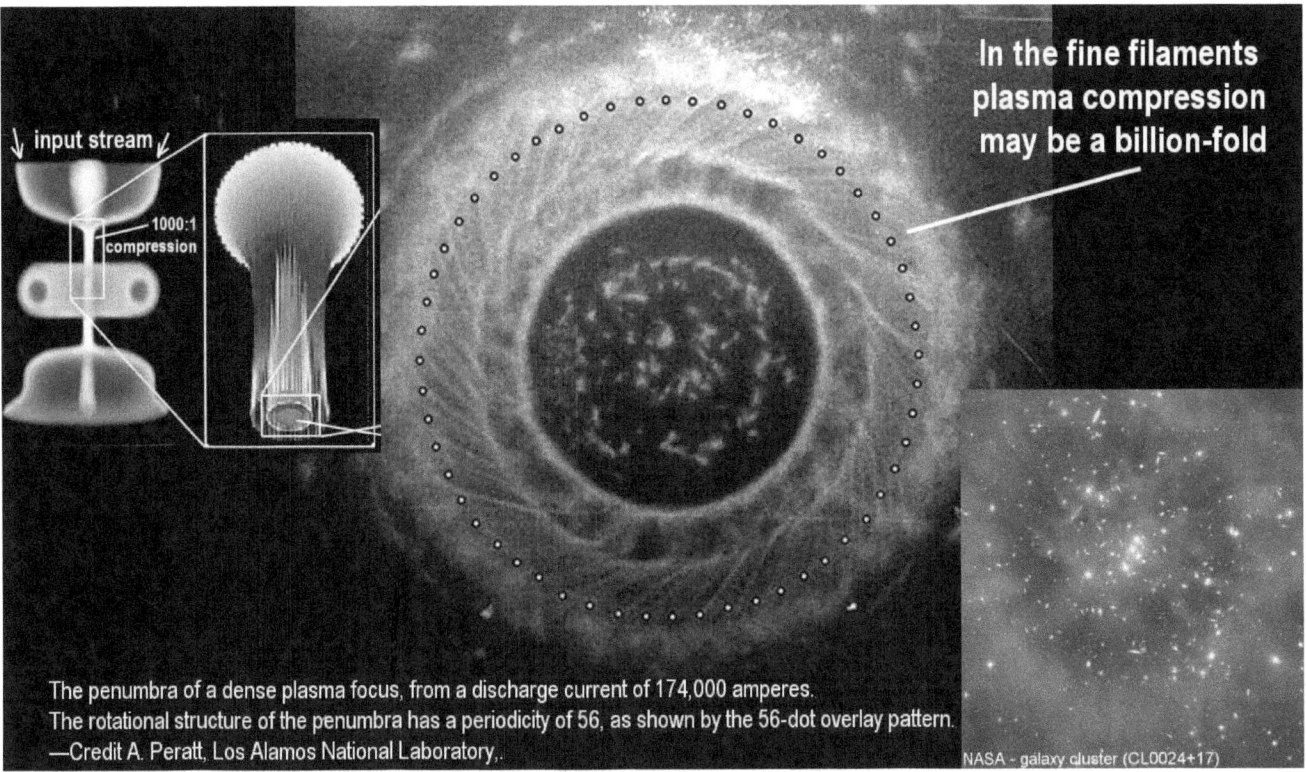

The penumbra of a dense plasma focus, from a discharge current of 174,000 amperes. The rotational structure of the penumbra has a periodicity of 56, as shown by the 56-dot overlay pattern.
—Credit A. Peratt, Los Alamos National Laboratory,.

NASA - galaxy cluster (CL0024+17)

In the fine filaments plasma compression may be a billion-fold

This means that by electromagnetic coupling, the rotation in the plasma stream is transmitted into the surface of the Sun that thereby rotates with the rotating magnetic field.

The Equator of the Sun rotates the fastest

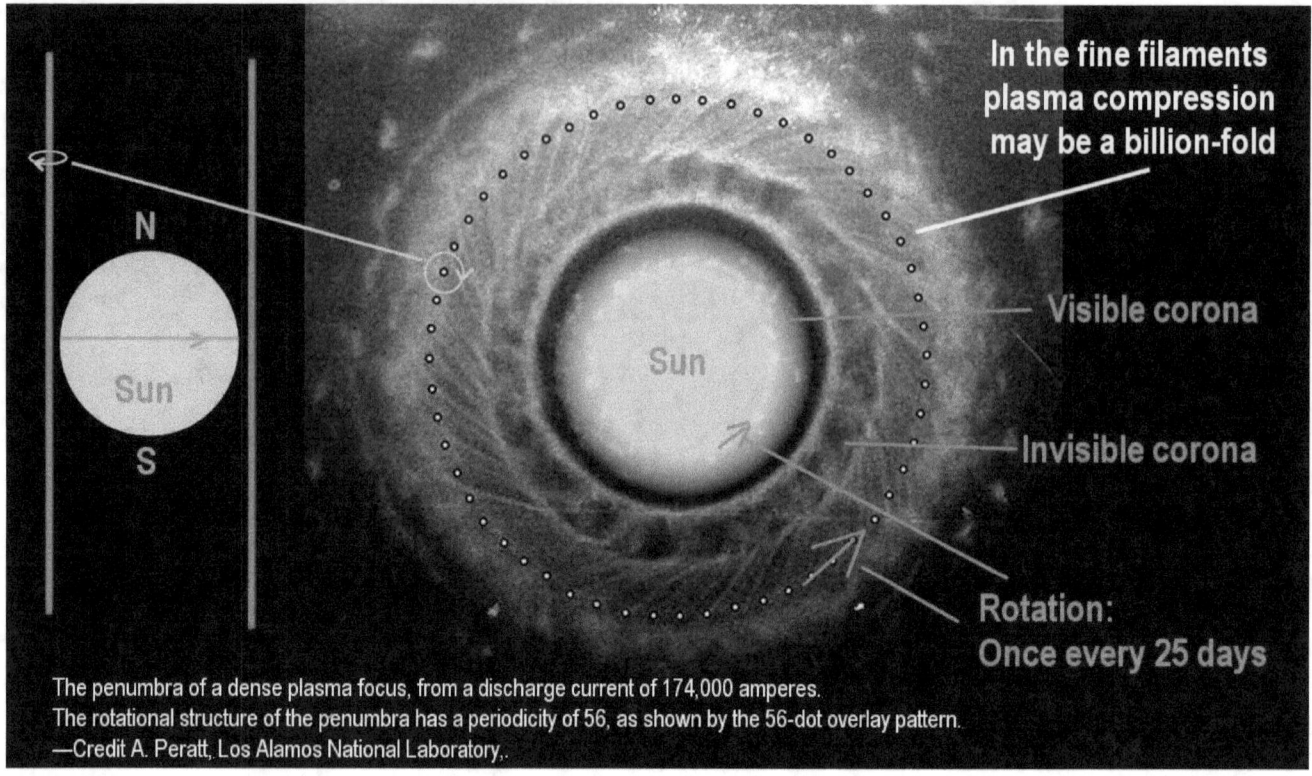

The penumbra of a dense plasma focus, from a discharge current of 174,000 amperes. The rotational structure of the penumbra has a periodicity of 56, as shown by the 56-dot overlay pattern.
—Credit A. Peratt, Los Alamos National Laboratory,.

Evidently, the resulting induction is the strongest, where the space to the Sun is the closest. By this principle, the Equator of the Sun rotates the fastest, and the polar regions the slowest, that are more weakly magnetically coupled.

At the equator a single rotation is completed in 25 days, and at the poles in 35 days. In the polar regions the low induction appears to actively slow the surface rotation, otherwise the entire Sun would likely spin at the same rotational speed, by internal coupling, as the gas planets do.

Perplexing as mechanistic phenomena

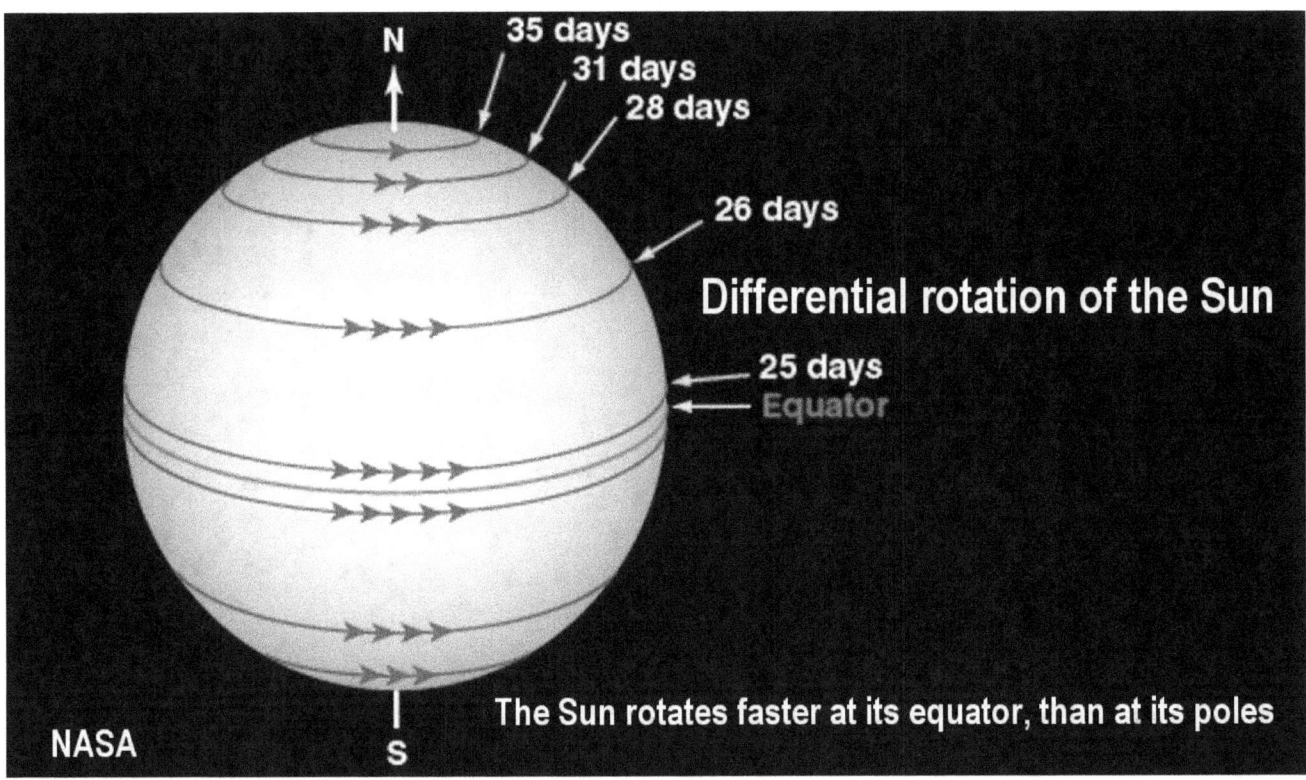

The differential rotation of the Sun is sometimes cited as the cause for the Sun's polar magnetic polarity reversal, and for the differential orientation of its magnetic field, both which are perplexing as mechanistic phenomena.

The 'language' of the real solar dynamics

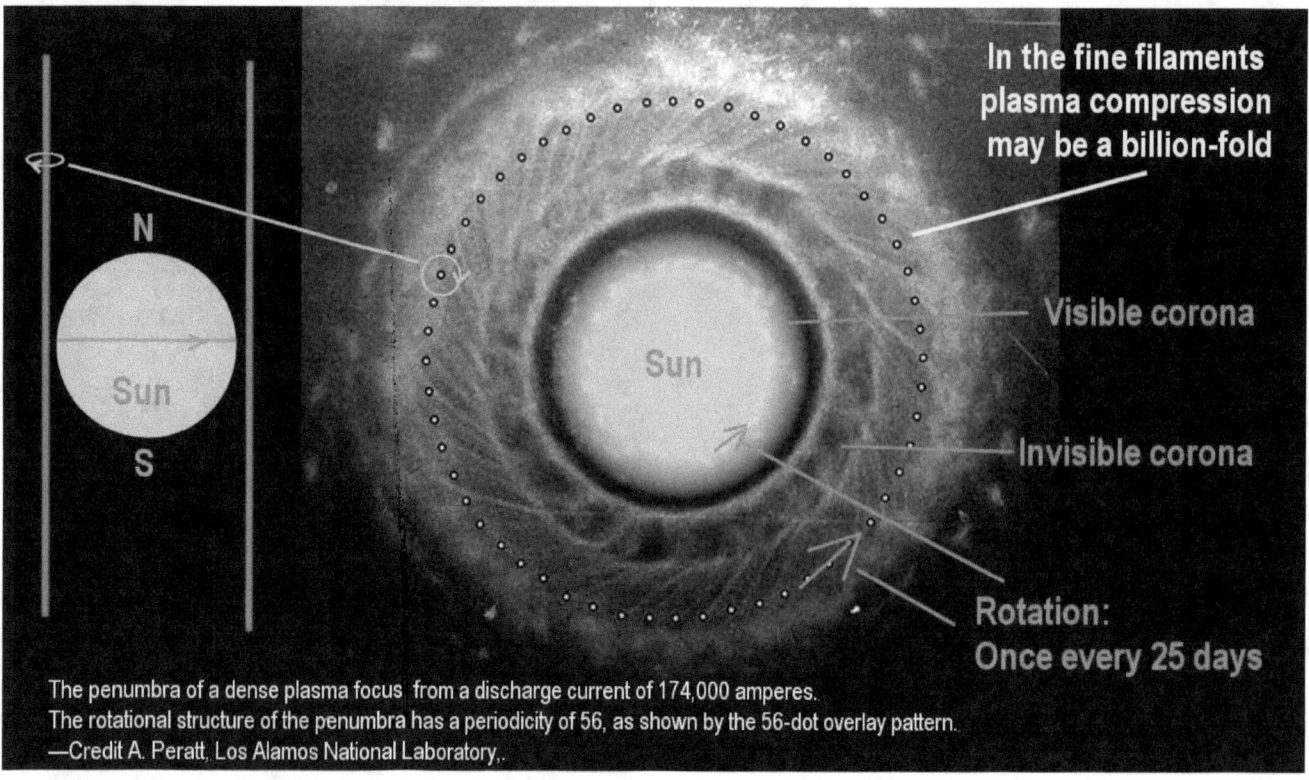

The penumbra of a dense plasma focus from a discharge current of 174,000 amperes.
The rotational structure of the penumbra has a periodicity of 56, as shown by the 56-dot overlay pattern.
—Credit A. Peratt, Los Alamos National Laboratory,.

But when we speak the 'language' of the real solar dynamics that are not mechanistic in nature, everything that we see expressed is governed by electrodynamics. Thus, speaking the correct 'language' solves many a perplexing puzzle

The 'language' of the plasma dynamics

The 'language' of the plasma dynamics even solves the paradox, why, when we look at the corona around the Sun, the differential between the various regions is not noticeable.

The same applies to the Sun in visible light

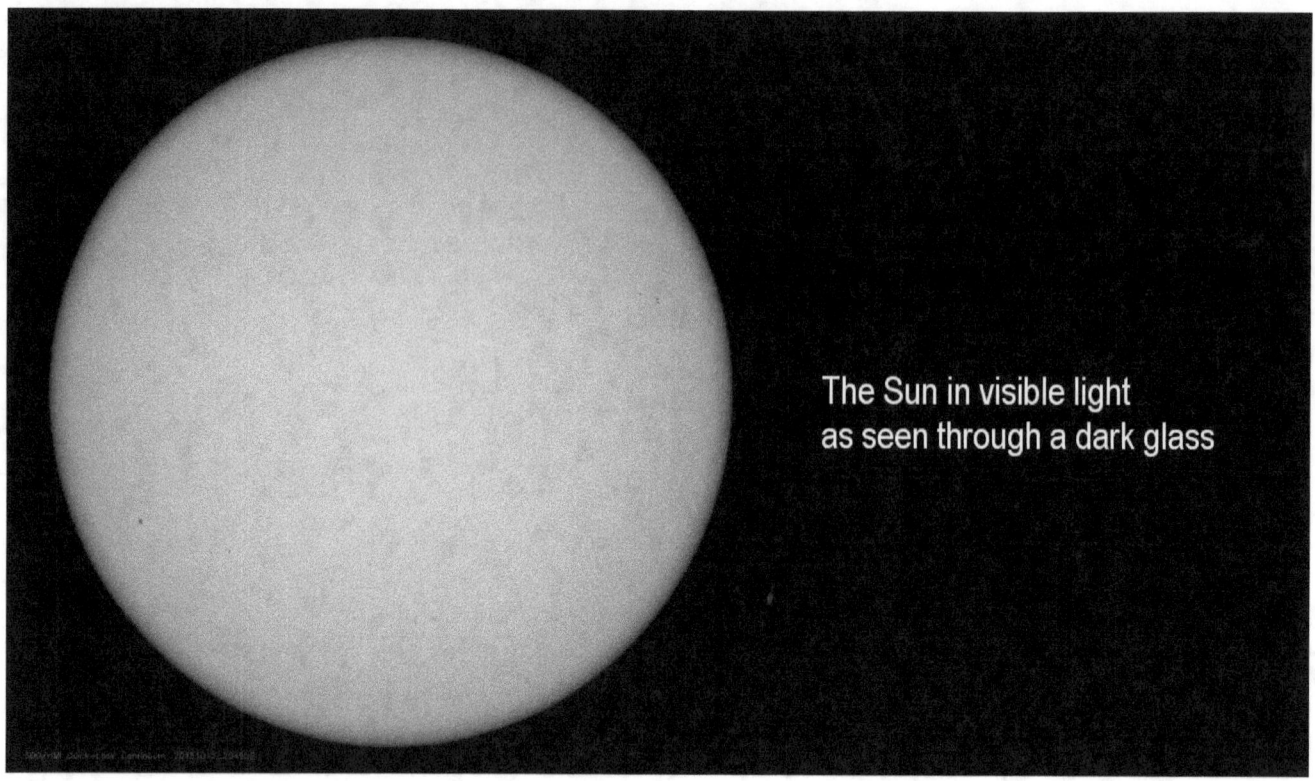

The Sun in visible light as seen through a dark glass

The same applies to the Sun in visible light.

Differential in solar activity

There, the differential in solar activity is only evident in erupting sunspots, and those hadn't even been noted until a few hundred years ago.

Sunspots remain largely a paradox

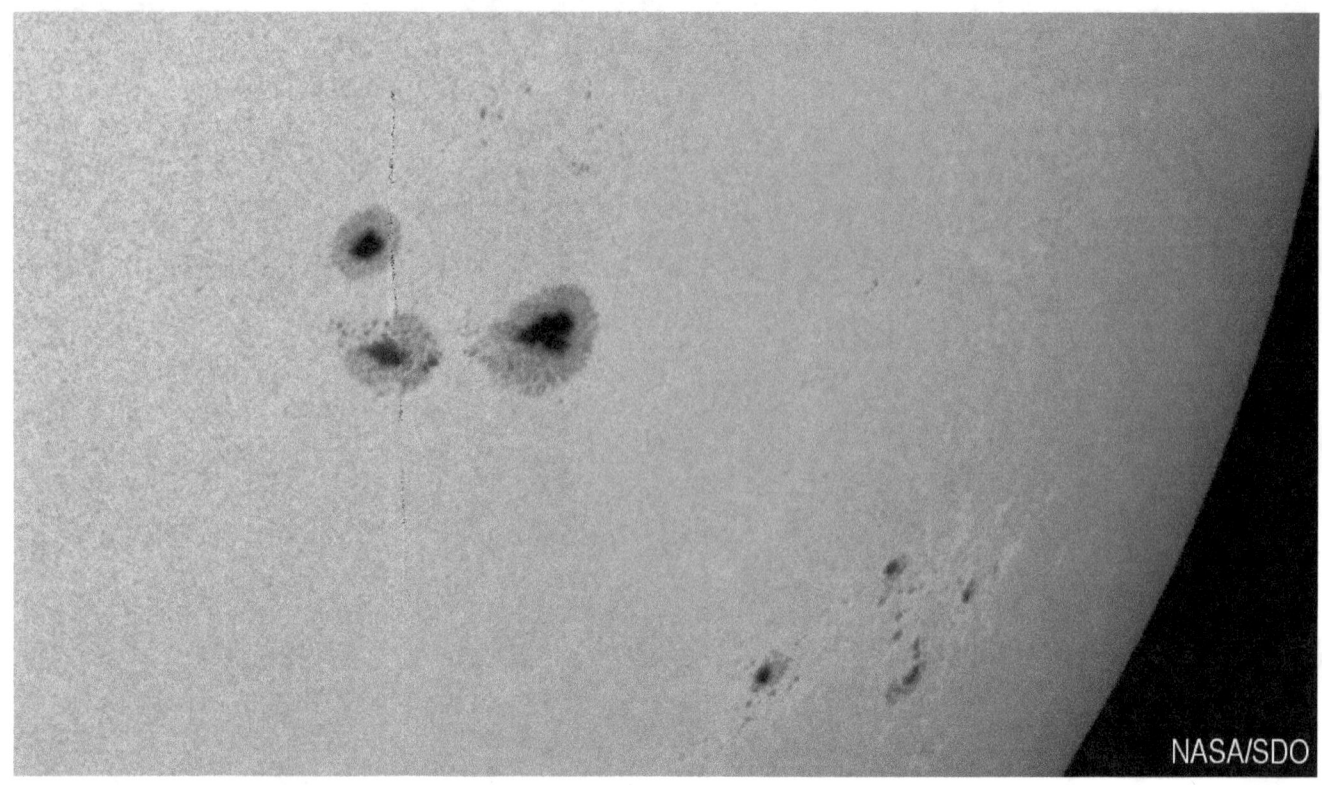

Ironically the sunspots remain largely a paradox to the present day.

The Sun's 'active' region

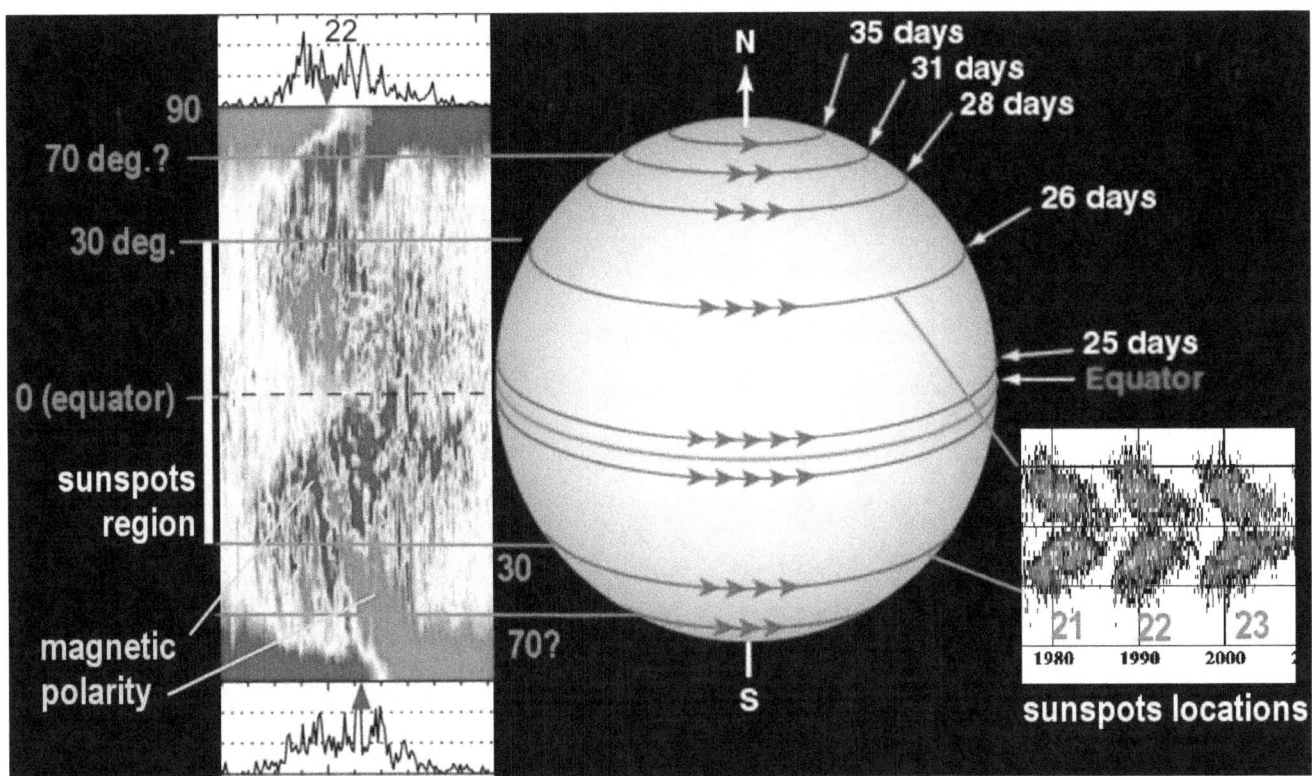

Nevertheless, the slight differential in the closeness of the surface of the Sun to the rotating filaments of the plasma stream that powers the Sun, in the entire band between the 30 degrees latitudes, becomes visibly apparent by this central band of the Sun being the Sun's 'active' region where most of the sunspots, loops, flairs, and so on, happen. And even this becomes only apparent when we look behind the scene with modern space based telescopes, such as the SOHO and SDO satellites, and begin to search for the drivers for what we see, because a lot of what we see there isn't really possible on the mechanistic platform.

The Sun's symmetric magnetic-field orientation

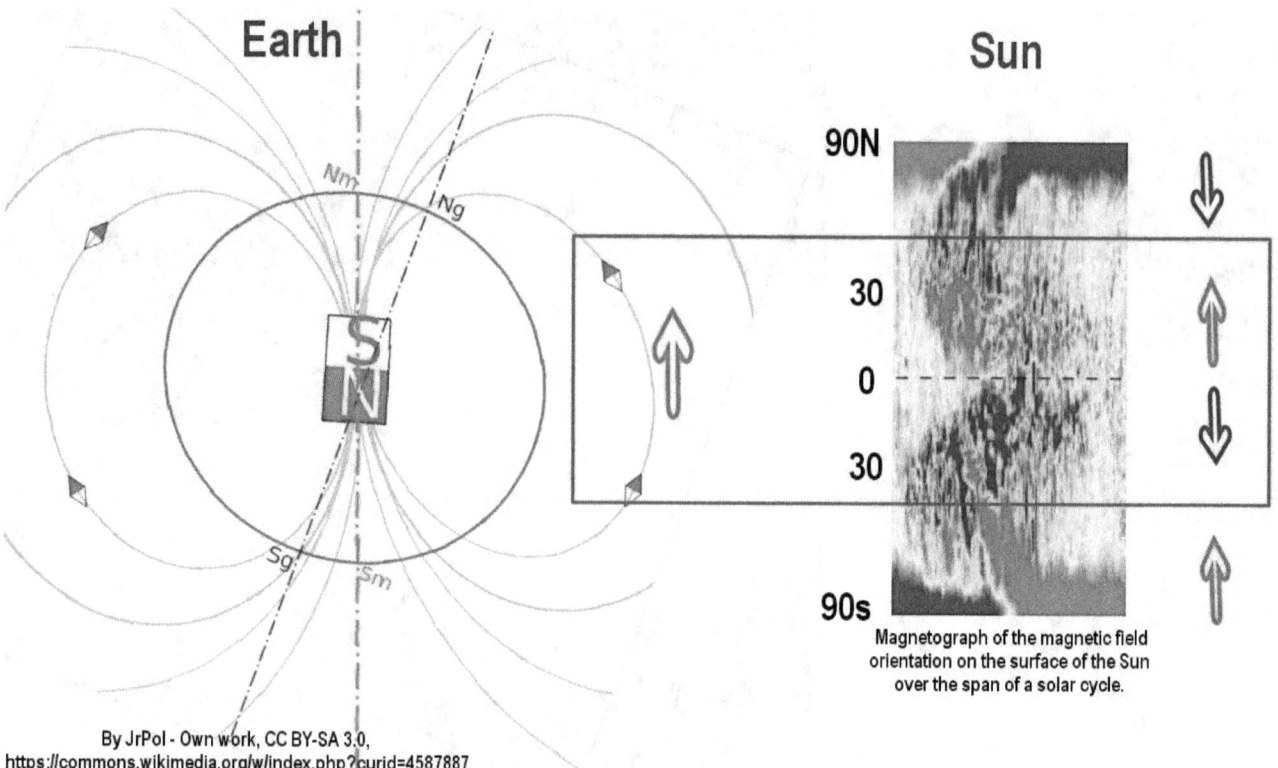

Magnetograph of the magnetic field orientation on the surface of the Sun over the span of a solar cycle.

By JrPol - Own work, CC BY-SA 3.0, https://commons.wikimedia.org/w/index.php?curid=4587887

One of the impossible aspects that shouldn't happen, is the Sun's symmetric magnetic-field orientation relative to its equator. What we see happening on the Sun is amazingly contrary to what we see on Earth.

➢ The Sun's magnetic field

```
┌─────────────────────────────┐
│   The Sun's magnetic field  │
│        like none other      │
└─────────────────────────────┘
```

The Sun's magnetic field

like none other

Opposite directions, symmetric to the equator

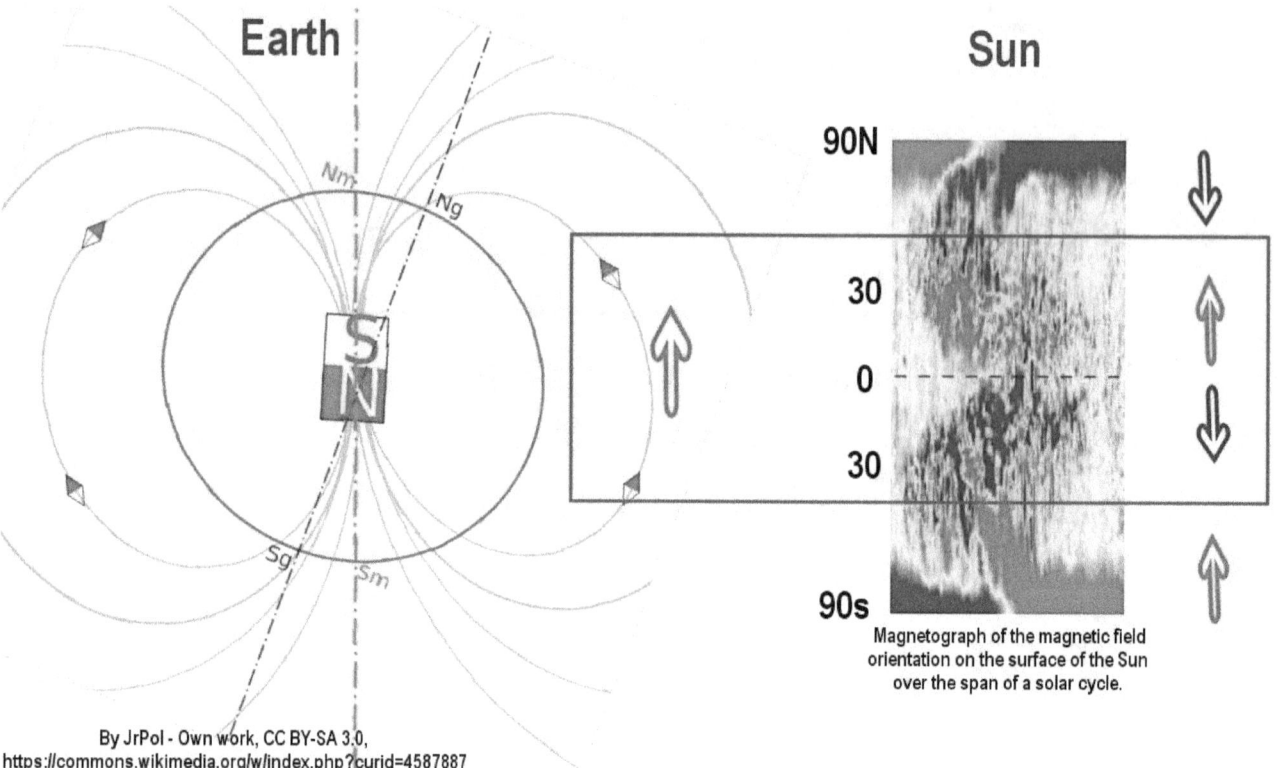

Magnetograph of the magnetic field orientation on the surface of the Sun over the span of a solar cycle.

By JrPol - Own work, CC BY-SA 3.0, https://commons.wikimedia.org/w/index.php?curid=4587887

On the Earth, the magnetic field that forms around it, is only oriented in one single direction; between North and South.

It is believed that a dynamo effect creates a central dipole field on Earth that has magnetic field lines flowing from pole to pole, so that the magnetic field lines all flow in the same direction. Consequently the compass points to the magnetic North Pole on Earth no matter where one may be located.

The Sun doesn't operate that way. The Sun has its magnetic field imposed onto it, and in a peculiar manner. On the Sun, we encounter two sets of magnetic fields that are oriented in opposite directions, symmetric to the equator.

The color coded magnetograph that is shown here, presents the changing magnetic field orientation as it has been detected on the surface of the Sun over the span of a solar cycle. The symmetry of the pattern divides the Sun into two symmetrically opposite hemispheres. What we see indicated here, wouldn't be possible on the Earth, and shouldn't be possible on the mechanistic Sun either, but it is totally real. The Sun speaks a different language as it were.

The dividing feature that separates the Sun into two magnetically opposite hemispheres is only possible in the plasma 'language'. This feature appears to be the existence external, symmetric magnetic fields that are separated by the heliospheric current sheet that is precisely aligned with the the equator.

heliospheric current sheet

heliospheric current sheet, is a 'thin' sheet of plasma that is as perfectly aligned with the equator of the Sun, as are the rings of Saturn, to Saturn's equator. Both structures form a thin disk, as thin and flat as a galaxy is in principle. Evidently the ring structures, the galactic structure, and the structure of the heliospheric current sheet, all share a common principle.

The Rings of Saturn, are at the planet's ecliptic

The Rings of Saturn, are at the planet's ecliptic. The moons of Saturn orbit closely aligned to this ecliptic, just as the planets of the solar system are aligned to the solar ecliptic.

In comparison, the heliospheric current sheet

The rings around Saturn are about 10 meters thick and extend outward as a perfect plain to roughly 80,000 kilometers.

In comparison, the heliospheric current sheet is a whopping 10,000 km thick near the Earth, and extends potentially across 15 billion kilometers of spacewhere it connects up with the distant heliosphere of plasma that surrounds the solar system.

The perfect alignment

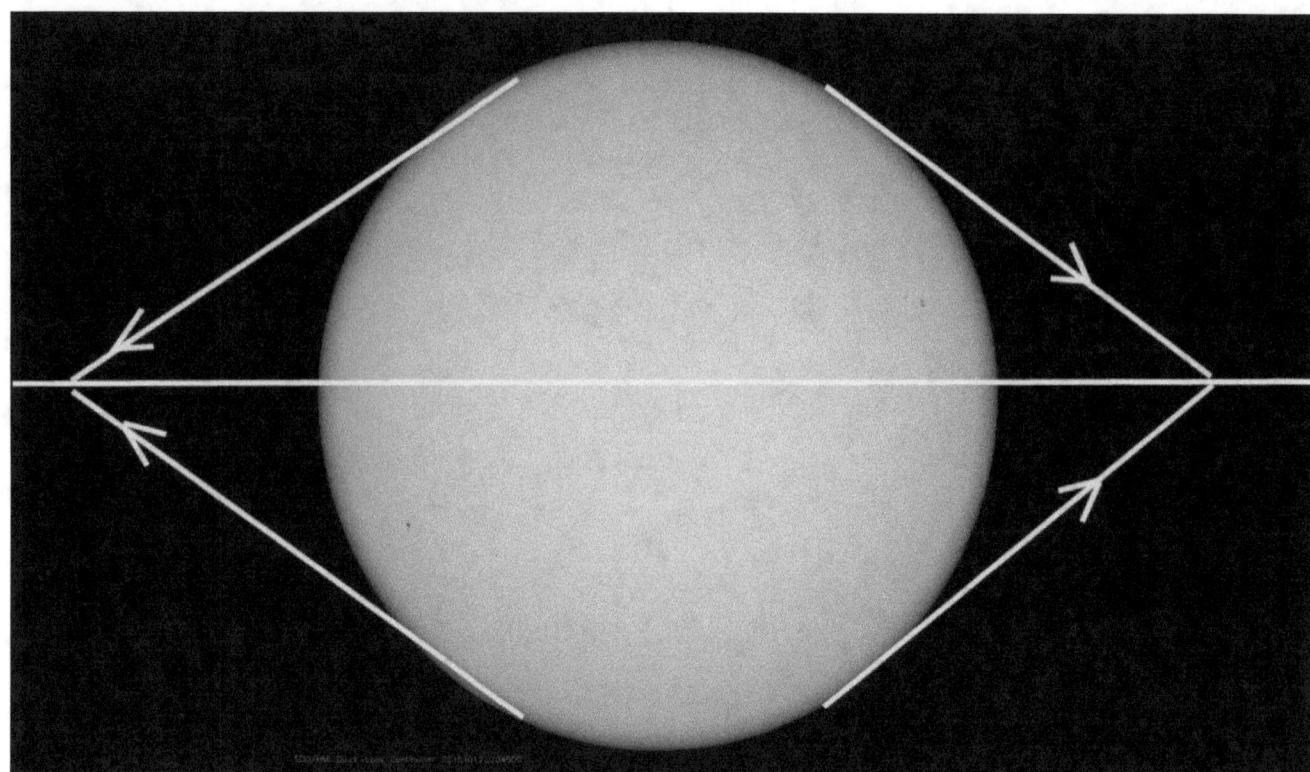

The perfect alignment of the heliospheric current sheet with the ecliptic is not by chance, but is evidently driven by the electric-force repulsion, which is the repulsion of electric elements of equal electric polarity. With the surface of the Sun being electron intense around the entire sphere, whatever plasma flows outward from the Sun is repelled from both sides equally, as both hemispheres are identical in electric potential. The current sheet is aligned thereby and is actively propelled outward.

To serve as a guide for magnetic fields

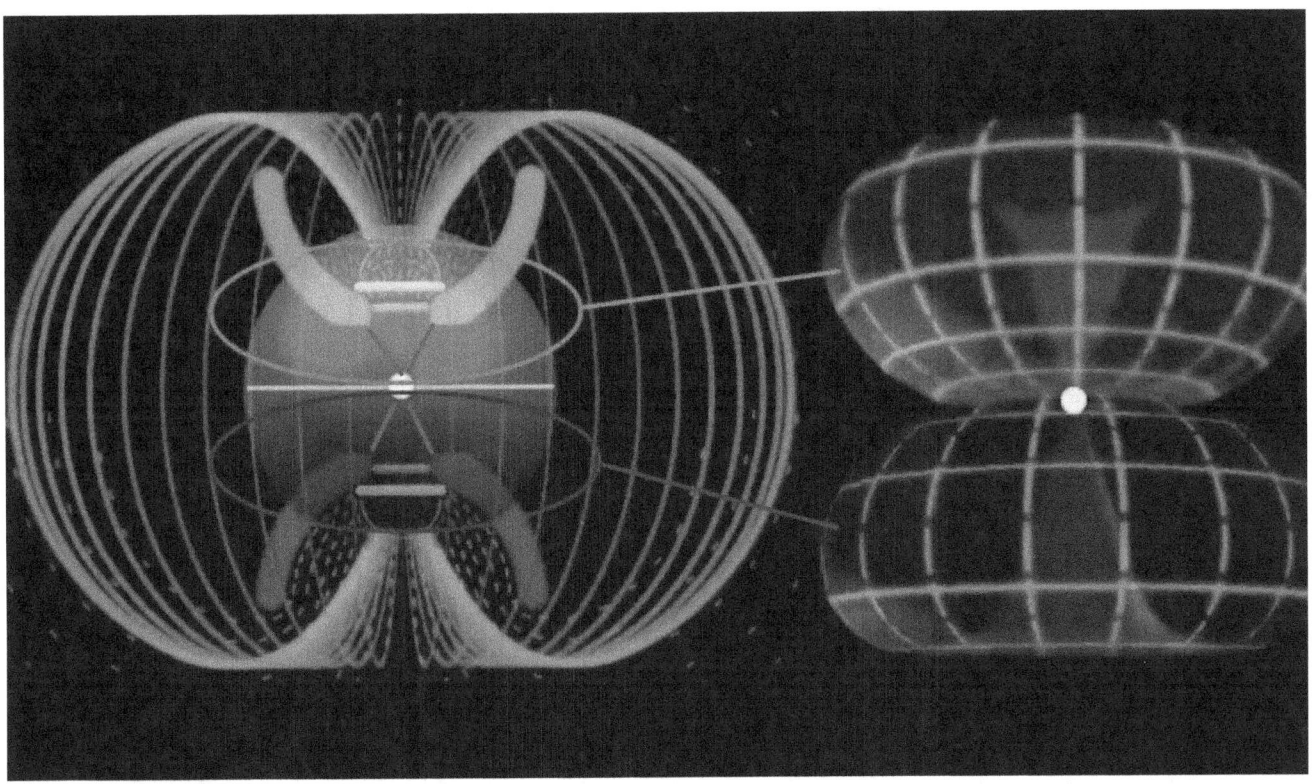

As I said, the heliospheric current sheet is aligned with the equator of the Sun and extends out from it along the plane of the ecliptic all the way to the heliosphere. This giant plasma sheet appears to serve as a guide for magnetic fields, a type of insulator that keeps the magnetic fields separated that evidently extend from each of the primer fields, each with an opposite magnetic polarity.

This has an amazing effect on the Sun. The effect is so powerful that the magnetic field lines in the Sun's hemispheres are able to reorient the flow of plasma loops, which flow from sunspots, into different directions in each hemisphere, according to the orientation of its field lines.

Sunspots are damaged areas

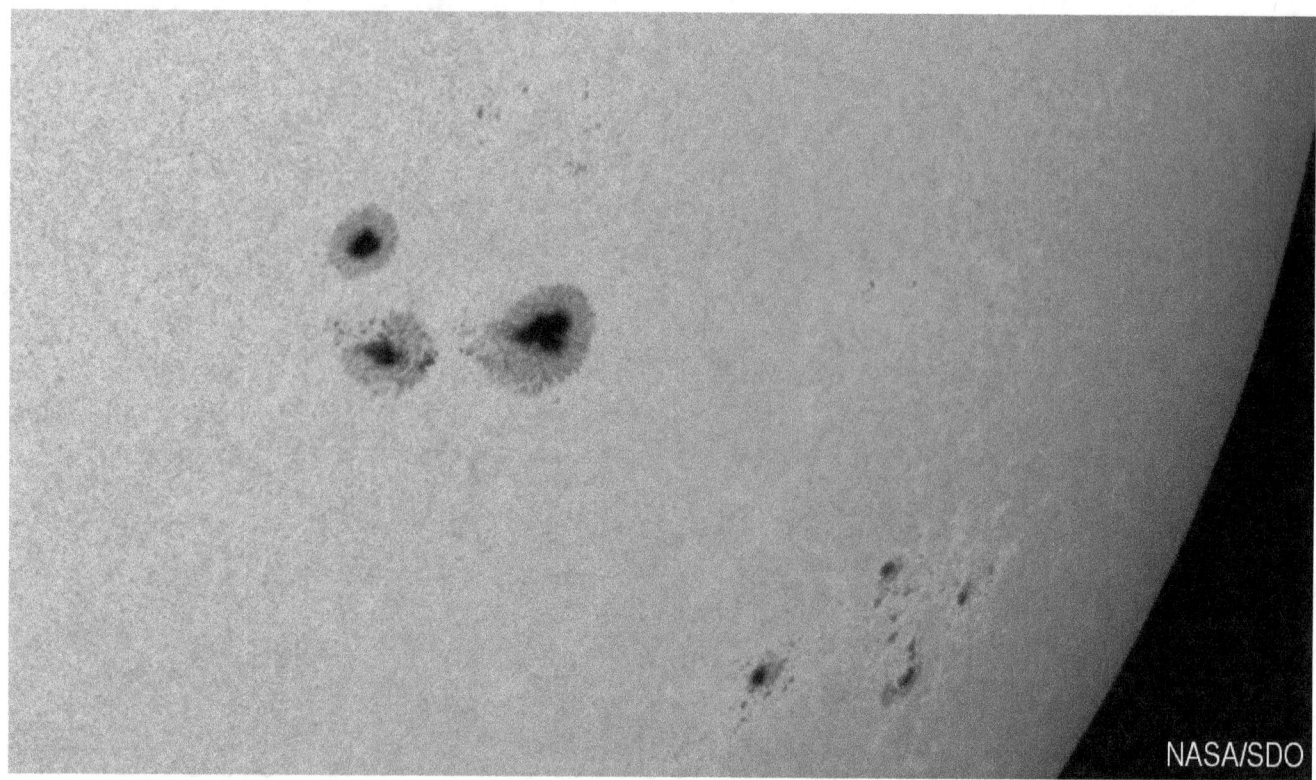

Sunspots are damaged areas on the surface of the Sun, that result from erupting plasma currents. If the currents are weak, the surface magnetic field bends the plasma stream back towards the surface of the Sun, according to the direction in which the field lines are oriented.

The eruption of sunspots creates loops

The eruption of sunspots thereby creates loops with a specific directionality, depending on in which direction the magnetic field bends the loop.

The eruption flow, and the re-entry flow

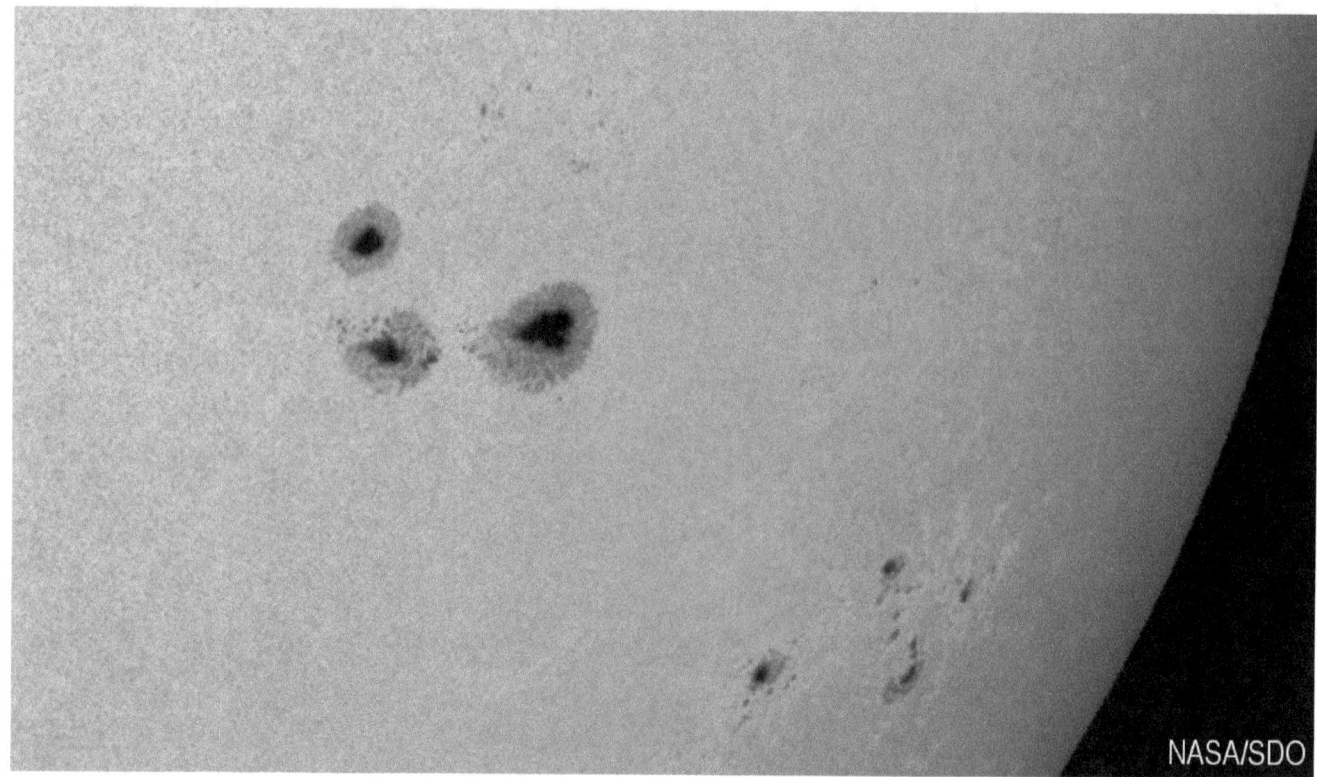

The plasma eruptions that occur on the Sun, when they are bent back into the Sun, typically result in pairs of closely spaced sunspots that are typically horizontally aligned. This means that the eruption flow, and the re-entry flow, typically create a pair of magnetized areas on the surface of the Sun, nearby the sunspots, but with opposite magnetic polarity. The relative position of the opposite polarities, to each other, indicates in which direction the magnetic field has bend the loop.

Let's look at sunspot 2700

For an example, let's look at sunspot 2700, as seen on February 27/2018.

When seen with a telescope, we see it as just a tiny spot on the surface. It is so tiny that it is almost lost in the sea of brilliance. But when one looks at the same area with a satellite's magnetic imager camera, one sees two oppositely polarized regions appearing, side by side, coded green and yellow in this example.

When one looks at the same spot in the high UV light band, one can even see the plasma loops happening that caused the spots. The orientation of the loops is typically perpendicular to the orientation of the magnetic field lines that bend the plasma flow.

The images shown here were created by NASA's SDO satellite.

Another example from May 24/2015

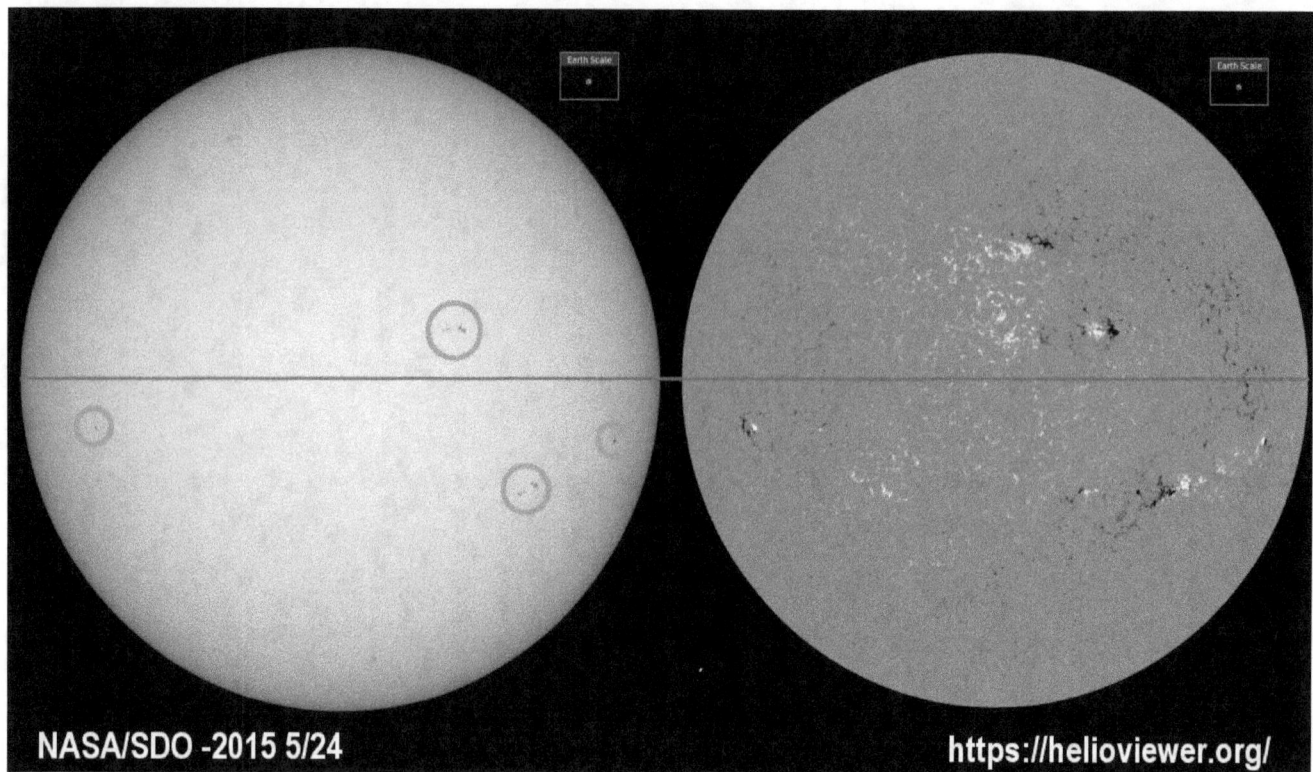

Now, lets look at another example from an earlier time, from May 24/2015. The image on the left, shows four sunspot groups in visible light. And the image on the right, shows the magnetic imaging of the same areas. Note that in the upper hemisphere, the black-colored polarity appears to the right of the white, while in the lower hemisphere, the black appears to the left of the white. This tells us that the magnetic field lines in the different hemispheres were oriented in opposite directions.

The differently oriented polarization proves

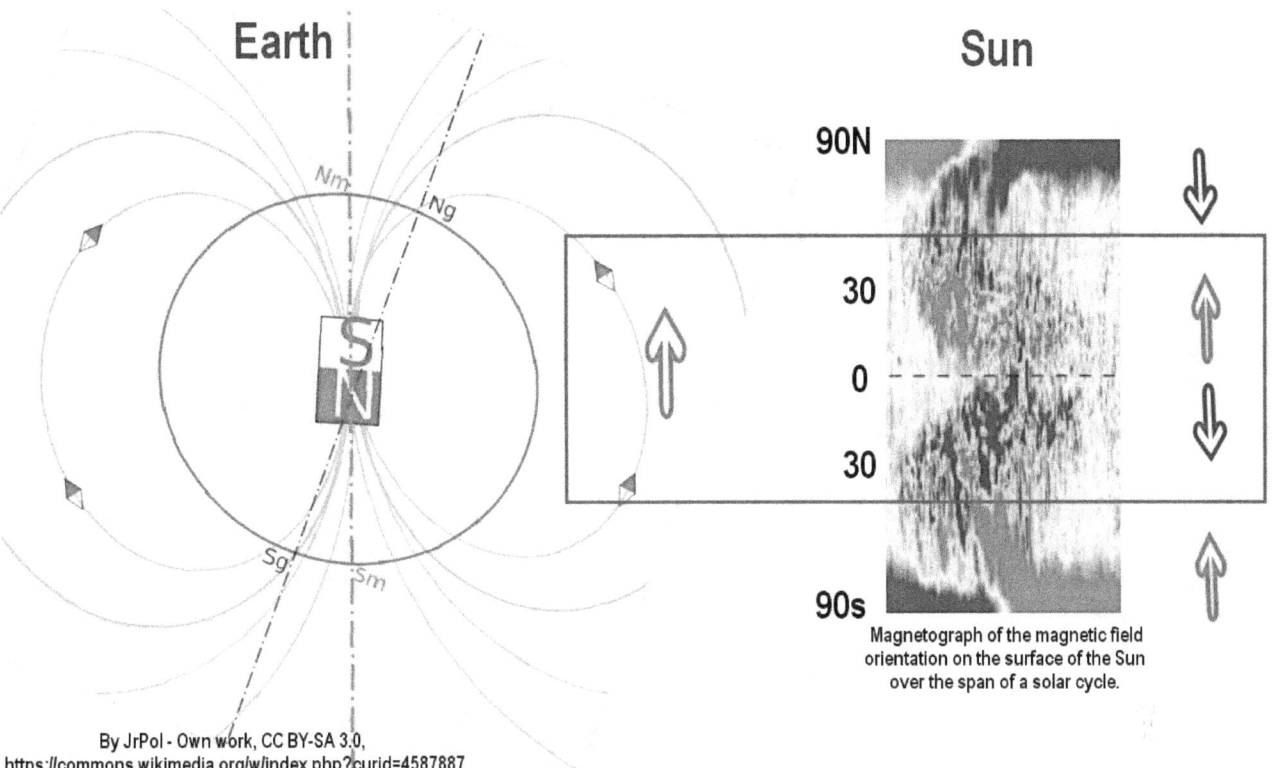

By JrPol - Own work, CC BY-SA 3.0,
https://commons.wikimedia.org/w/index.php?curid=4587887

Magnetograph of the magnetic field orientation on the surface of the Sun over the span of a solar cycle.

The differently oriented polarization proves rather graphically that the Sun's northern and southern hemispheres are magnetically symmetric, rather than homogenous as they are on the Earth.

Proof that the Sun's magnetic field is imposed

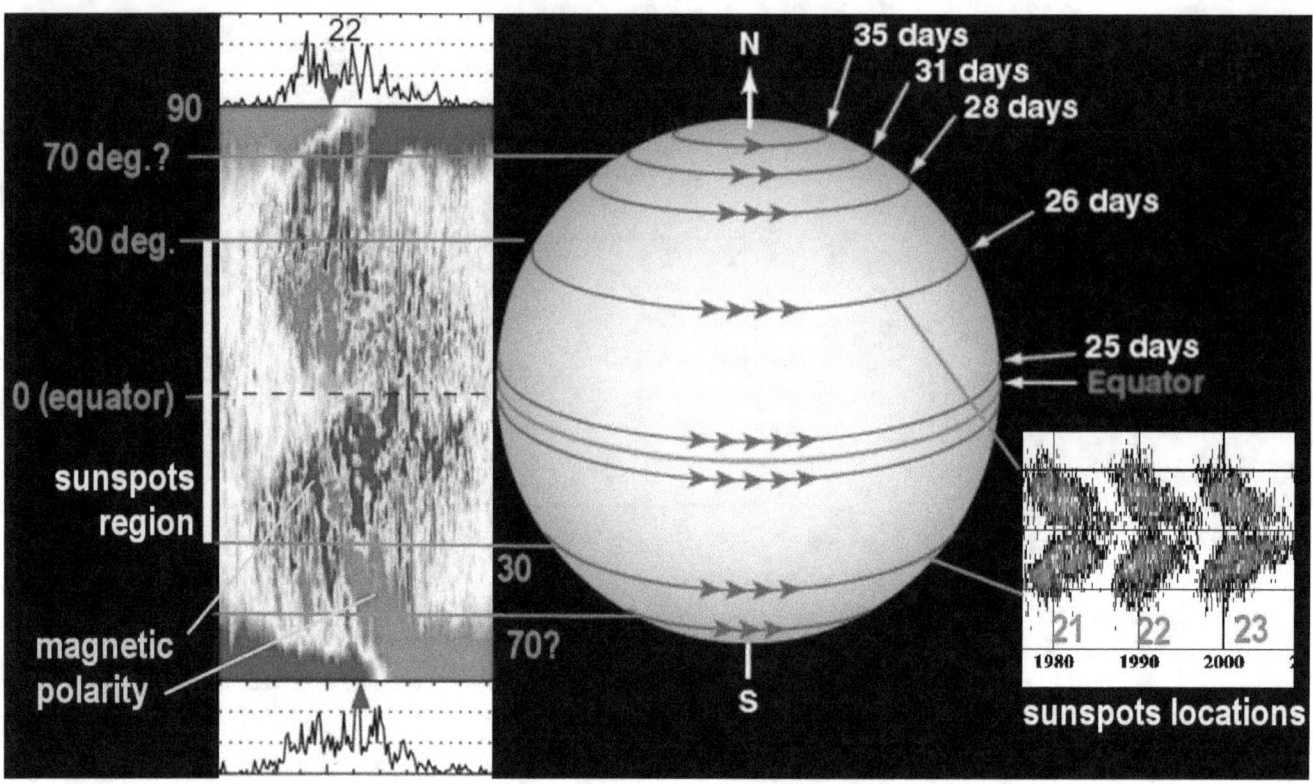

The polarity differential is a significant proof that the Sun's magnetic field is imposed onto it from the outside, by the magnetic action of different primer fields.

Loop structures that don't produce sunspots

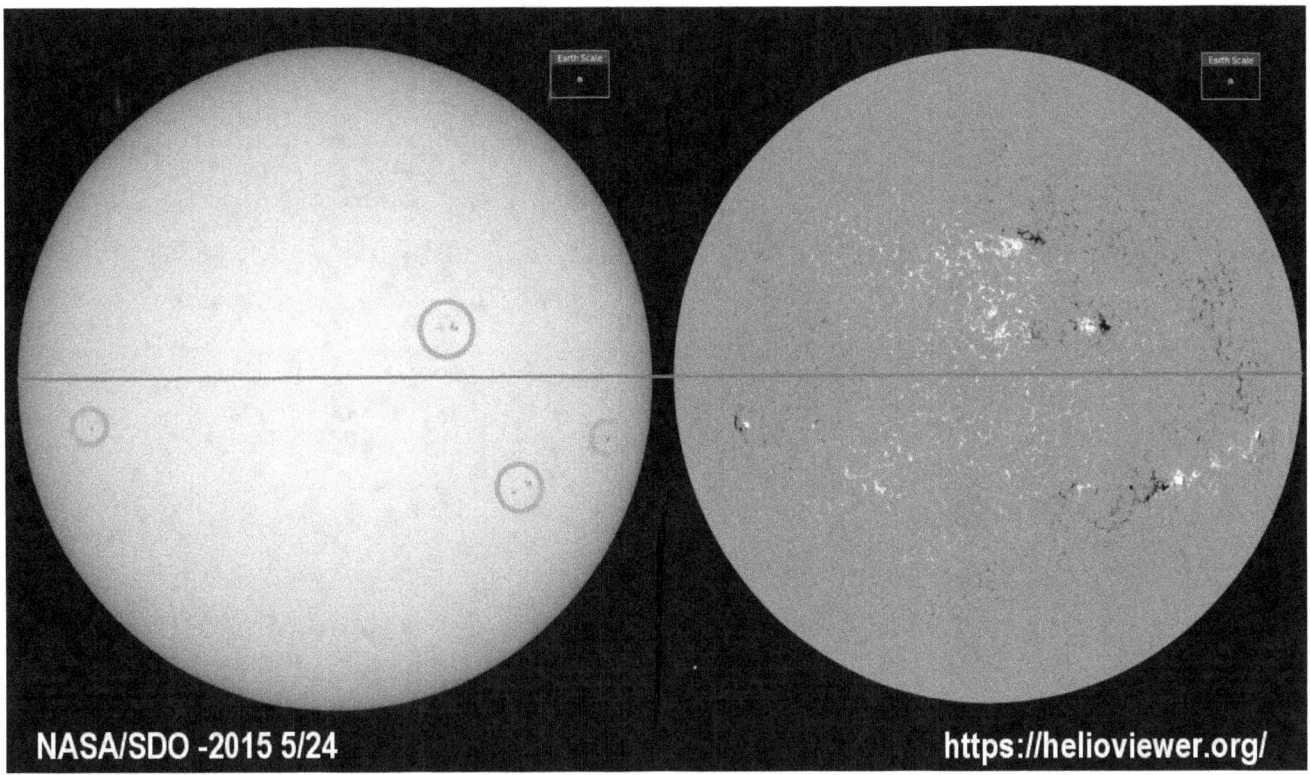

It also proves something else. It proves that not all loop structures on the Sun are strong enough to create sunspots. The uppermost magnetic spot doesn't have a sunspot associated with it. It proves that sunspots are extreme events, and that lesser loop structures can exist that don't produce sunspots.

Loops associated with the magnetic spots

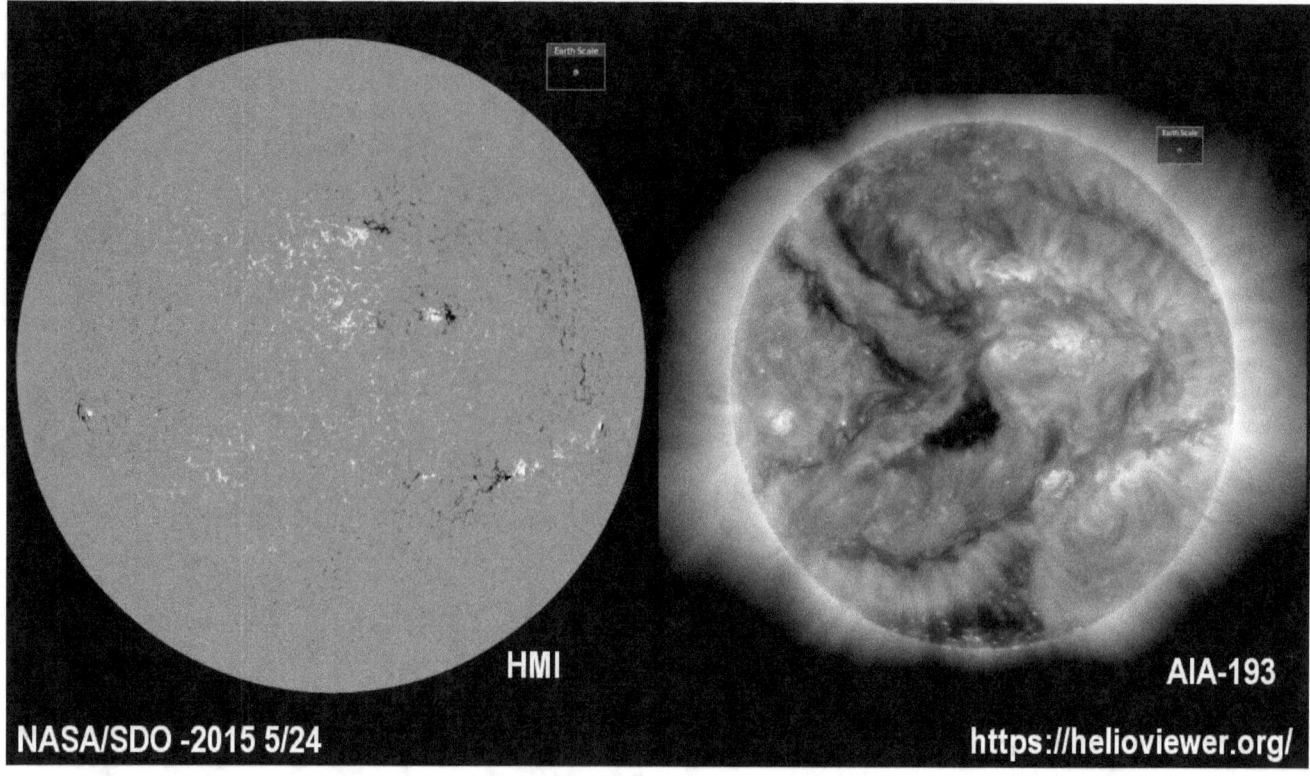

In this UV image, on the right, the plasma loops associated with the magnetic spots are faintly visible.

Plasma loops are occur horizontally

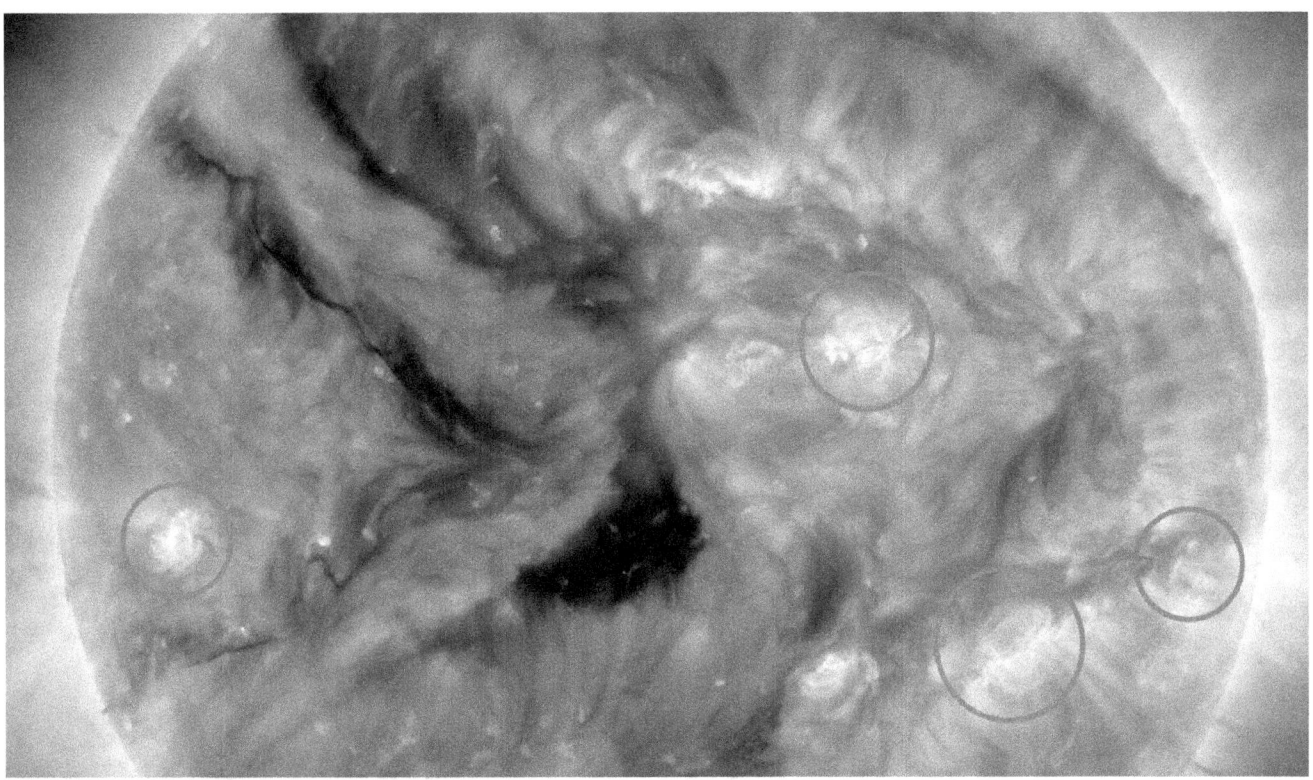

And in the more expanded view, here, its becomes clearly recognizable that the plasma loops are all occur essentially horizontally as one would expect them to be when being forced by magnetic fields extending from the equator. This effect is also seen at the upper loop that didn't cause a sunspot.

Sunspots from the previous solar cycle

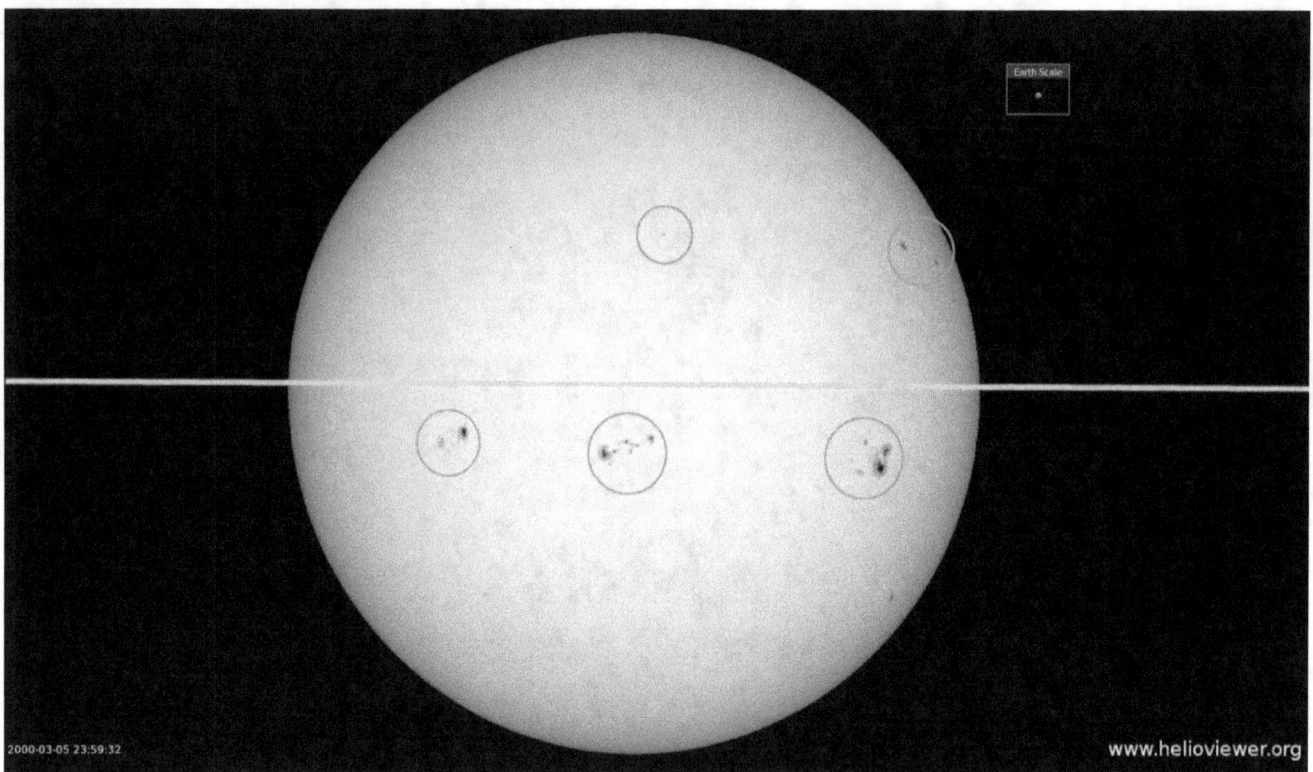

Now, let's look at another set of sunspots from the previous solar cycle, from March 2000. We see a string of sunspots extremely close to the equator, and a tiny one on top in the middle, and another one on the right.

Because the Sun's magnetic polarity flips

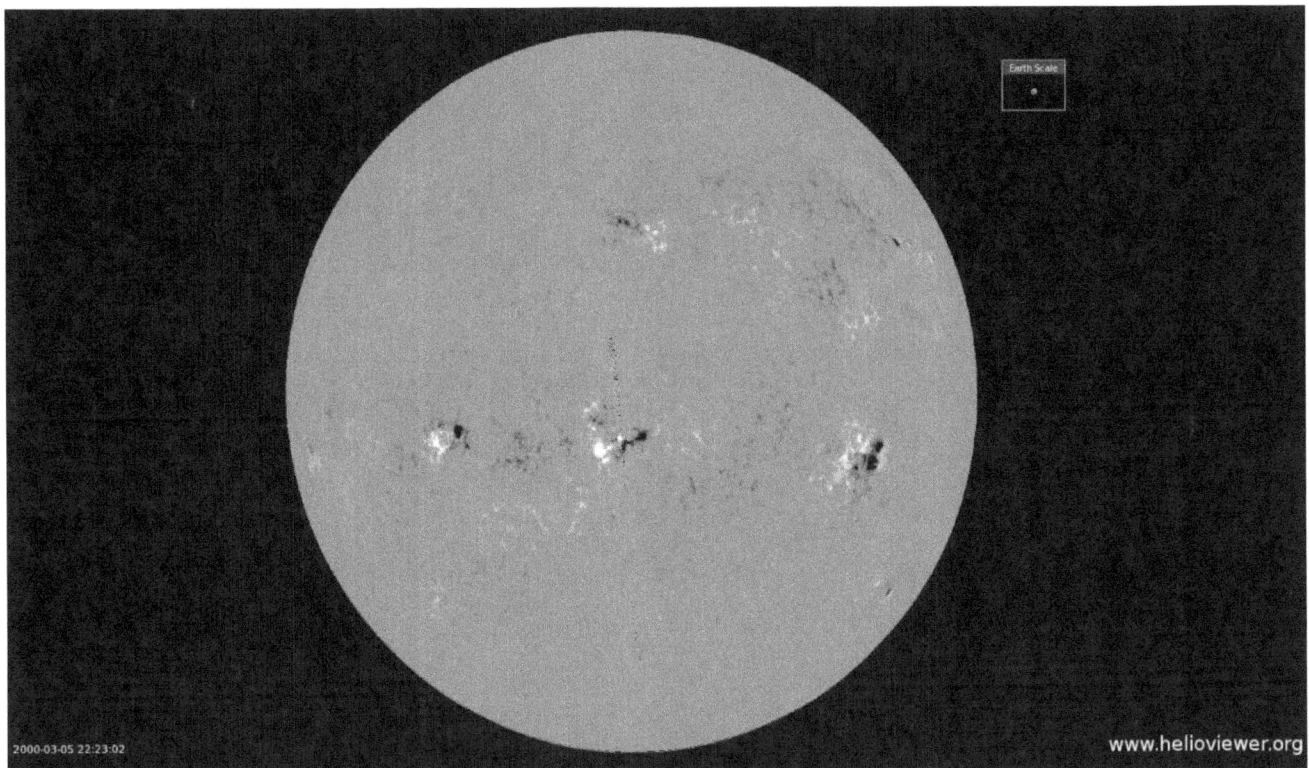

Because the Sun's magnetic polarity flips with every solar cycle, we see the polarity differentiation reversed. In the previous image we saw the white shadow on the left in the upper hemisphere. Now we see the white shadow on the right. This means that the shadow is now cast by the opposite magnetic orientation. The same reversal is also visible in the lower hemisphere, as it should be, because the solar magnetic polarity reversal flips the field lines in both hemispheres simultaneously, into their opposite orientation.

The symmetric magnetic field reversal proves

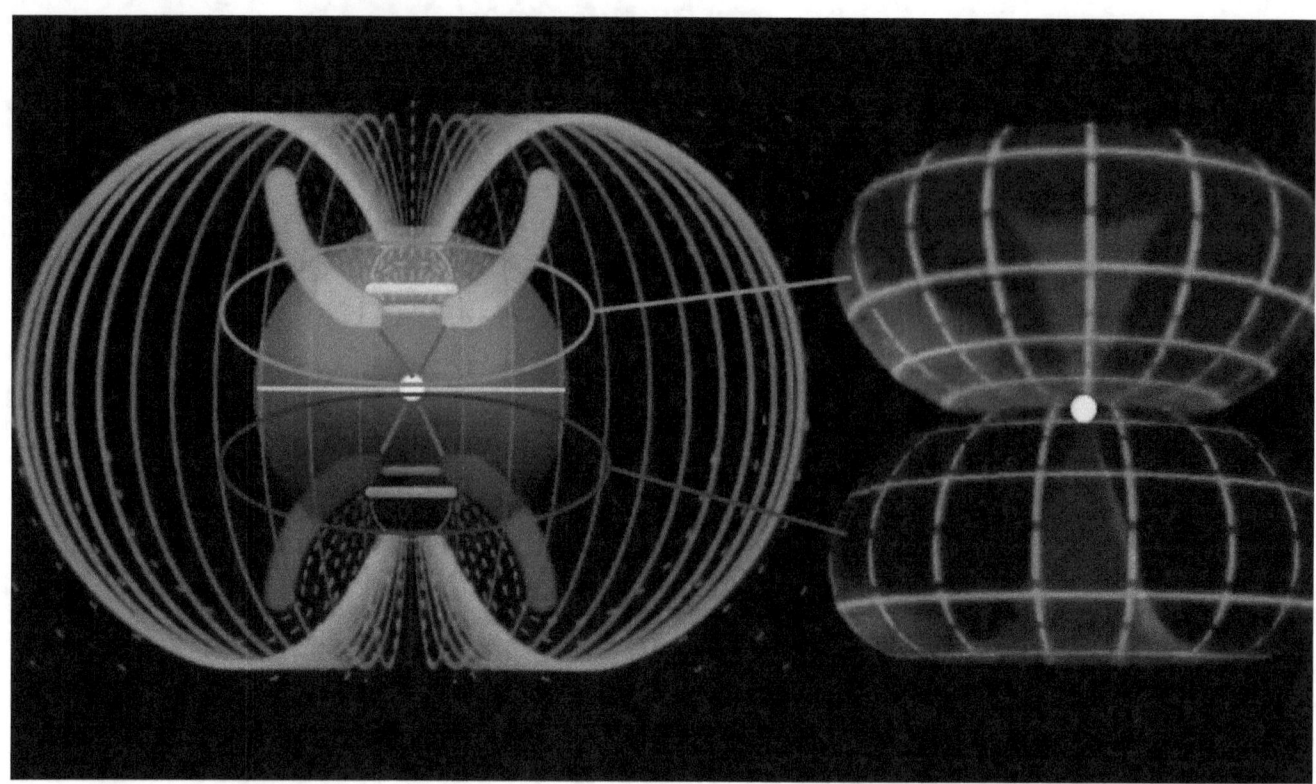

The symmetric magnetic field reversal proves that the primer fields, which impose their polarity, are very real, and are deeply connected with the most intimate functions of the Sun.

➢ The 22-year cycle

> The 22-year cycle with magnetic polarity reversal

The 22-year cycle with magnetic polarity reversal

How is the polarization reversal generated?

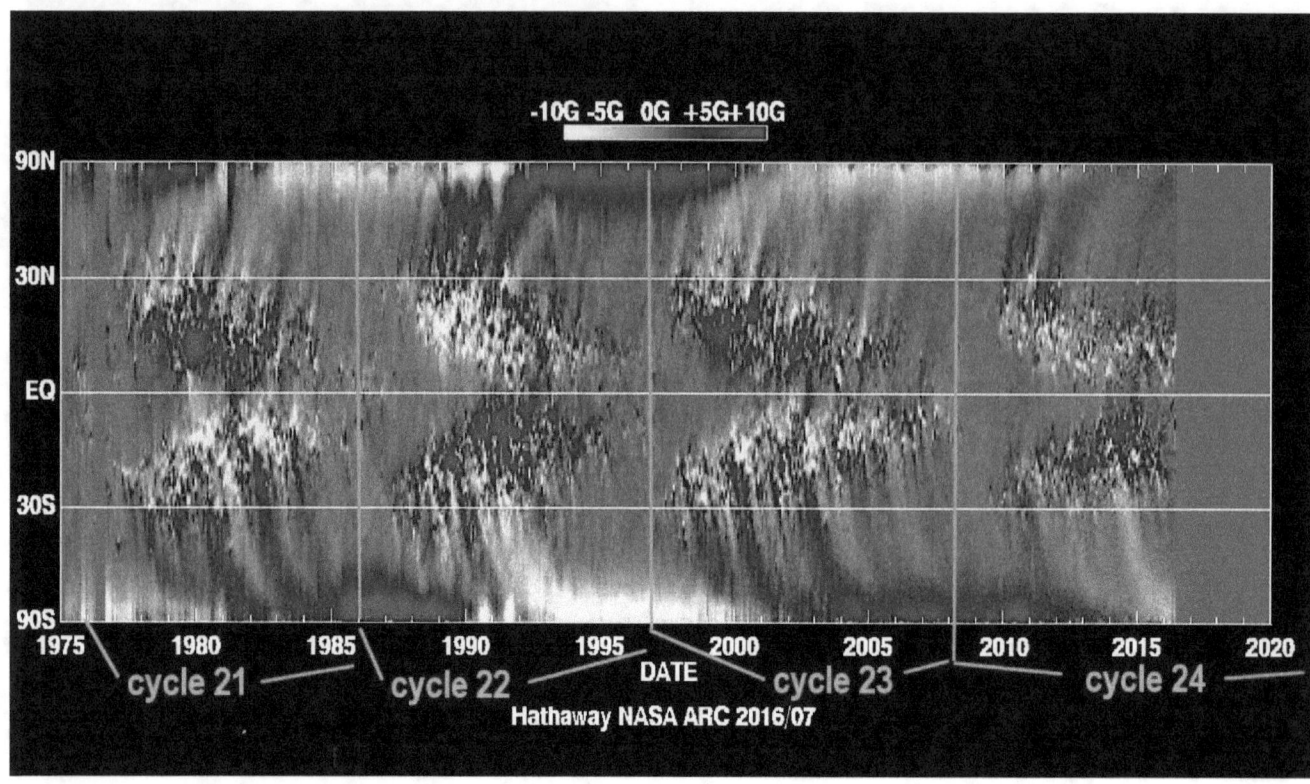

It has long been recognized that the Sun reverses all of its magnetic fields to their opposite polarization in a 11-year cycle, and back again in the next cycle, for a full cycle time of 22 years. The magneto-graphic illustration shown here indicates how the various regions of the Sun had been magnetically polarized over the spans of the last 45 years.

But how is the polarization reversal generated? What controls it? What controls the timing?

The answer is amazing

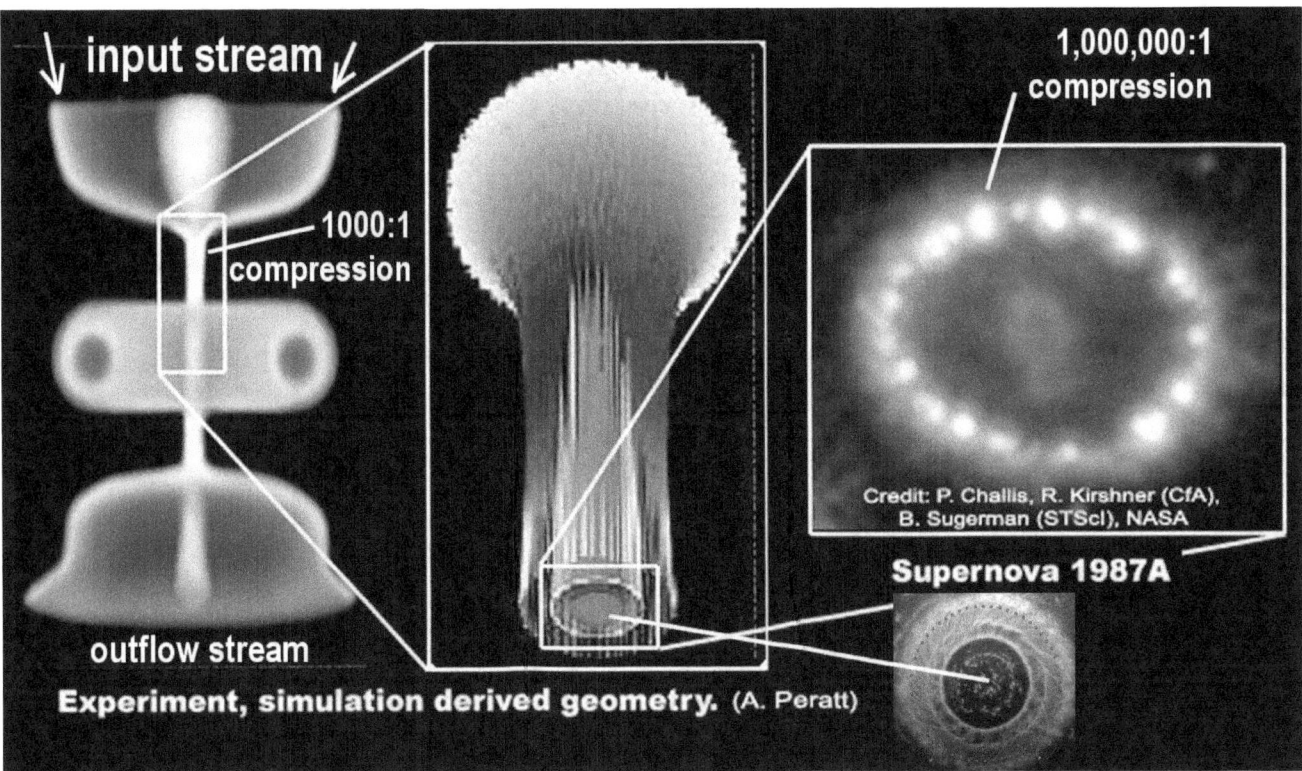

The answer is amazing. For this we need to go back to the high-energy discharge experiment by Anthony Peratt at the Los Alamos National Laboratory.

Differential creates a magnetic-field polarity

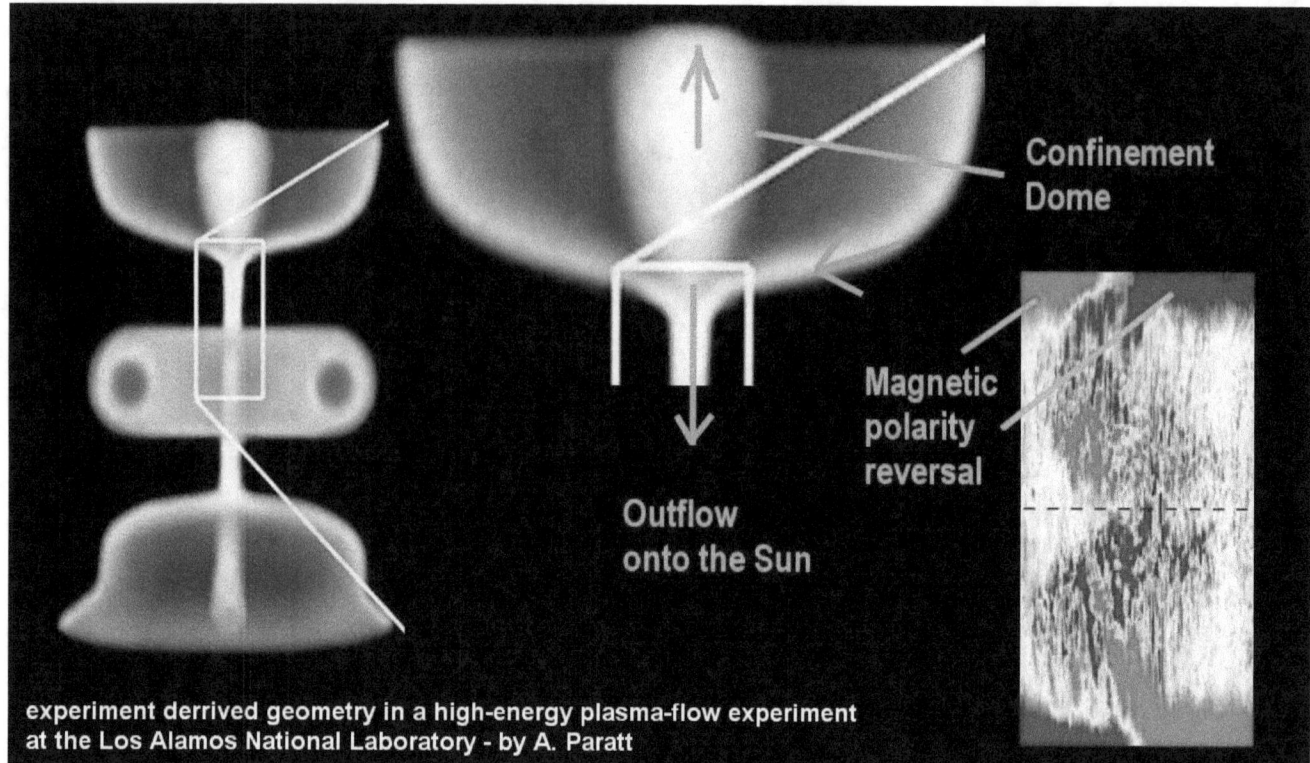

experiment derrived geometry in a high-energy plasma-flow experiment at the Los Alamos National Laboratory - by A. Paratt

The plasma structure that formed in the high-energy experiment, created a large confinement dome. The plasma pressure that builds up under the dome appears to be directed downward and increases the plasma outflow into the Sun. The differential creates a corresponding magnetic-field polarity.

However, when the pressure builds up in the confinement dome to such an intensity that it exceeds the strength of the containment field, the dome ruptures at the top, at its weakest point. When this happens, some of the built-up pressure in the dome vents into space in the opposite direction of flow. The change in balance, together with the magnetic disconnection at the top of the dome, generates a strong magnetic field with opposite polarity. The opposite polarity, thereby, becomes the dominant polarity from this point on. The polarity is maintained for as long as the outflow from the dome continues, or is forced to change by external causes.

Inversely, when the pressure in the dome drops, and the containment is re-established, or is re-established by external causes, the original magnetic polarity becomes dominant again. The magnetic field then flips again to its former orientation.

By this dynamic principle, a plasma stream flowing in the same direction, can create an oscillation of alternately oriented magnetic fields. This type of principle is evidently the cause for the cyclical polarity reversal that we see evident on the Sun.

Magnetic reversal in the polar regions of the Sun

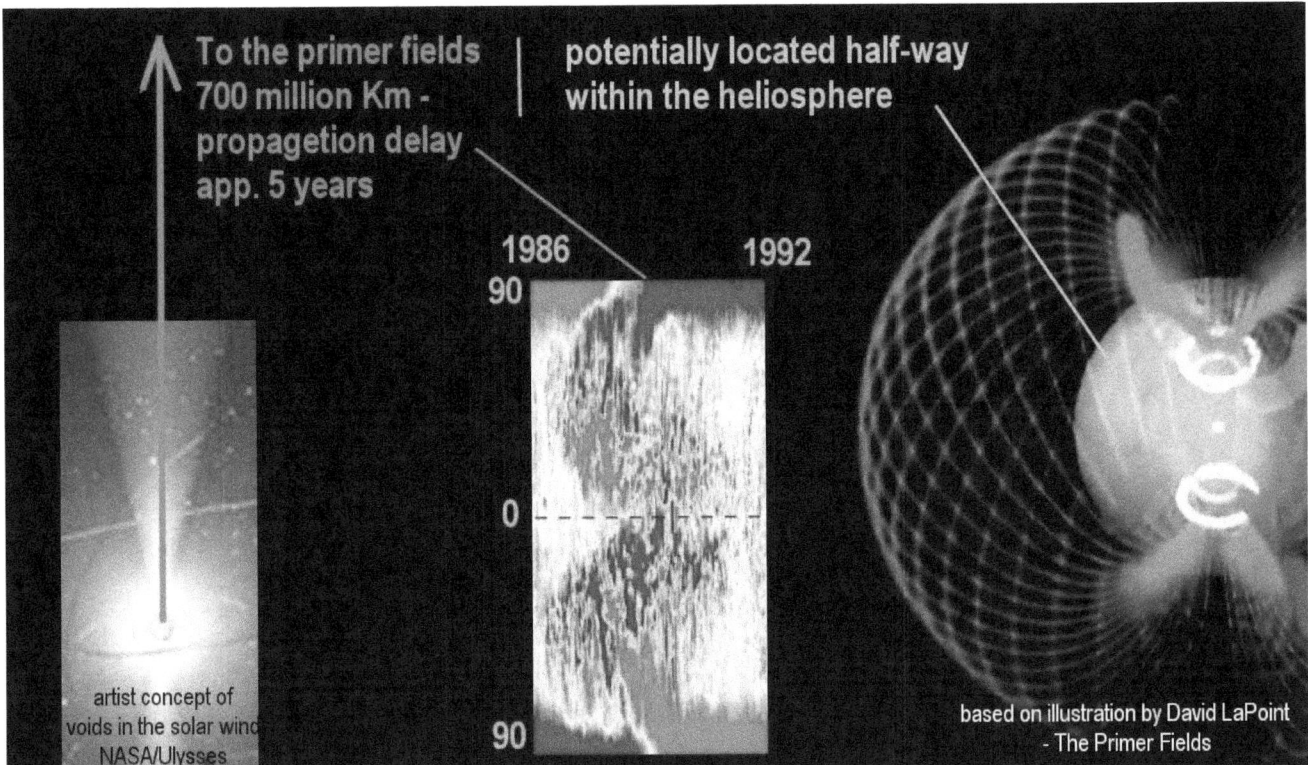

However, the polarity reversal isn't quite as simple as that. The magneto-graphic illustration, that shows the changing polarity in the regions of the Sun, from pole to pole, over the span of a solar cycle, shows that the magnetic reversal in the polar regions of the Sun, occurs late in the solar cycle, almost 5 years late. This delay creates a puzzle in mechanistic cosmology. It shouldn't be happening. The delay is theorized to be caused by magnetic material slowly drifting from the active regions to the poles. Can this be so?

In plasma language the delay is recognized

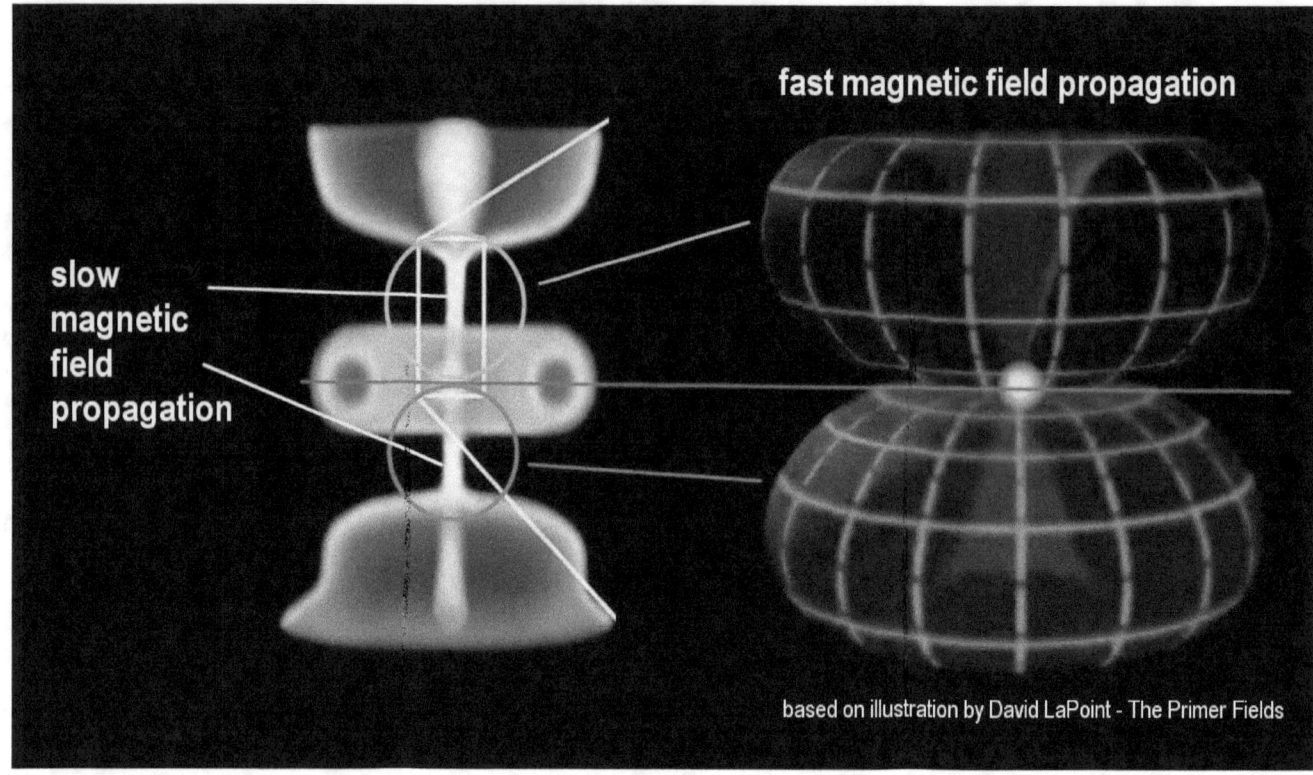

In plasma language the delay is recognized as the result of the natural differential of the magnetic field propagation caused by different media. In empty space, magnetic fields propagate at the speed of light. In dense plasma streams the propagation speed is determined by the density of the plasma, which is typically the plasma's pressure-propagation speed.

The Sun's distance to the primer fields

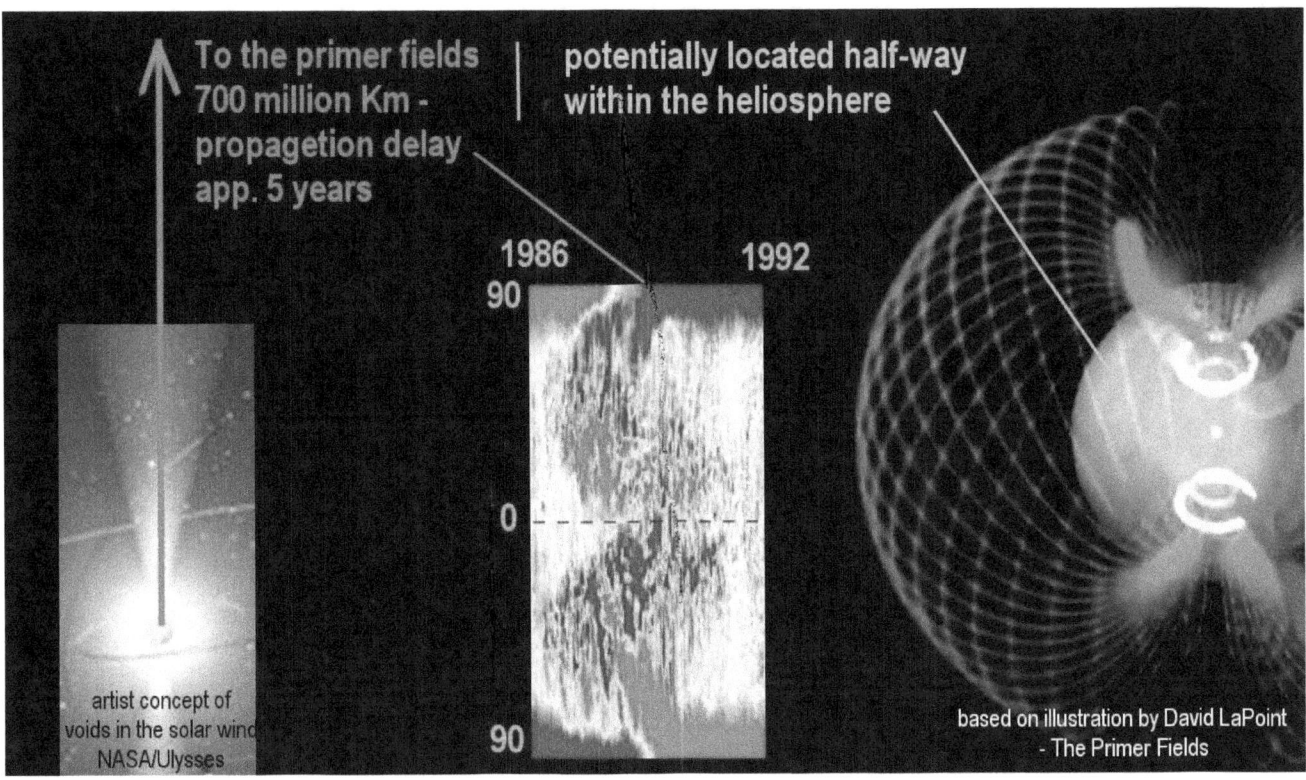

If one considers that the primer fields are potentially located half-way within the heliosphere, the Sun's distance to the primer fields is roughly 7 billion kilometers. The 5-year delay over this distance adds up to a propagation speed of 45 km/sec, which is well within the range of what is reasonably possible.

This means that the polarity reversal delay is not an enigma in plasma language.

For most of the Sun, the polarity reversal is not delayed

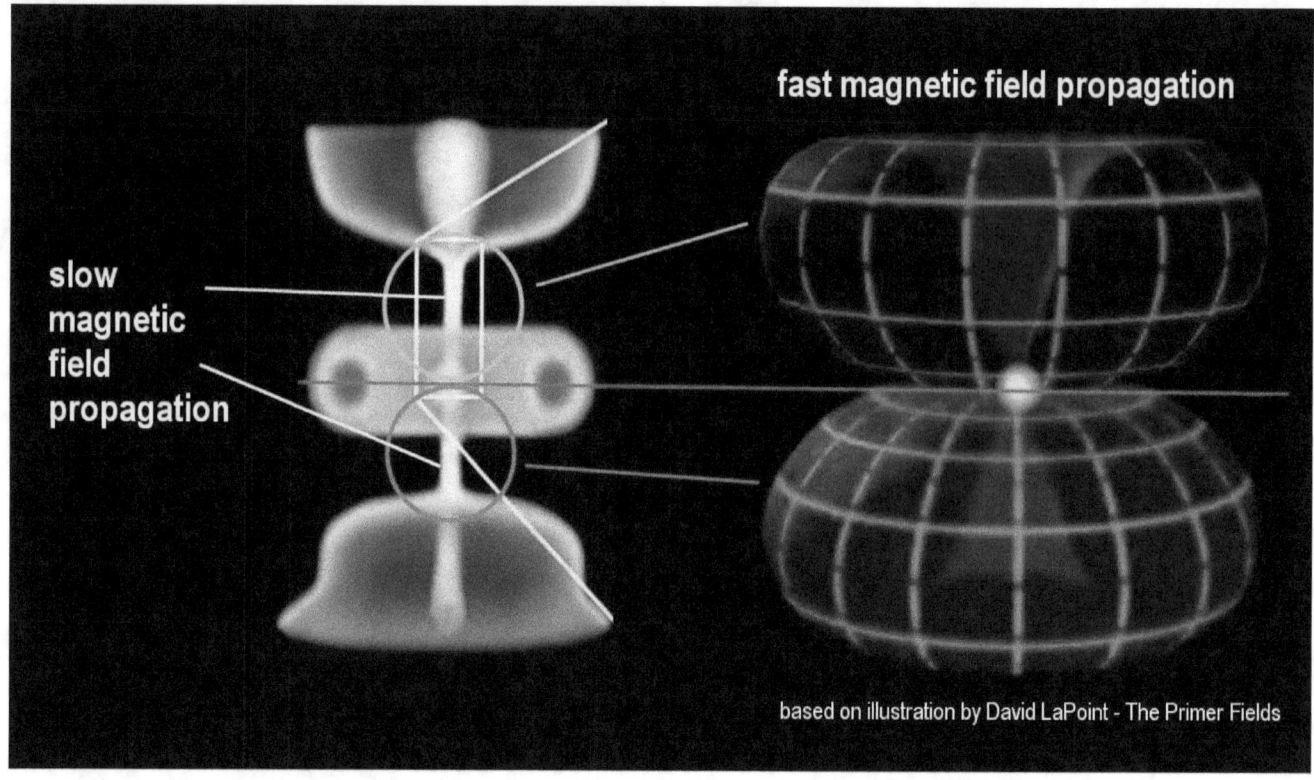

For most of the Sun, the polarity reversal is not delayed, because the magnetic field lines that flow outside of the plasma stream, propagate near the speed of light. The fast fields evidently don't reach all the way down to the poles. The poles would be dominated by the high-density plasma flow.

The slow plasma flow, in the high-density plasma stream, in comparison with the magnetic propagation speed that affects the central region of the Sun, is also responsible for the shape of the sunspot cycles as we see them evident increasingly.

The reason why the solar activity peaks

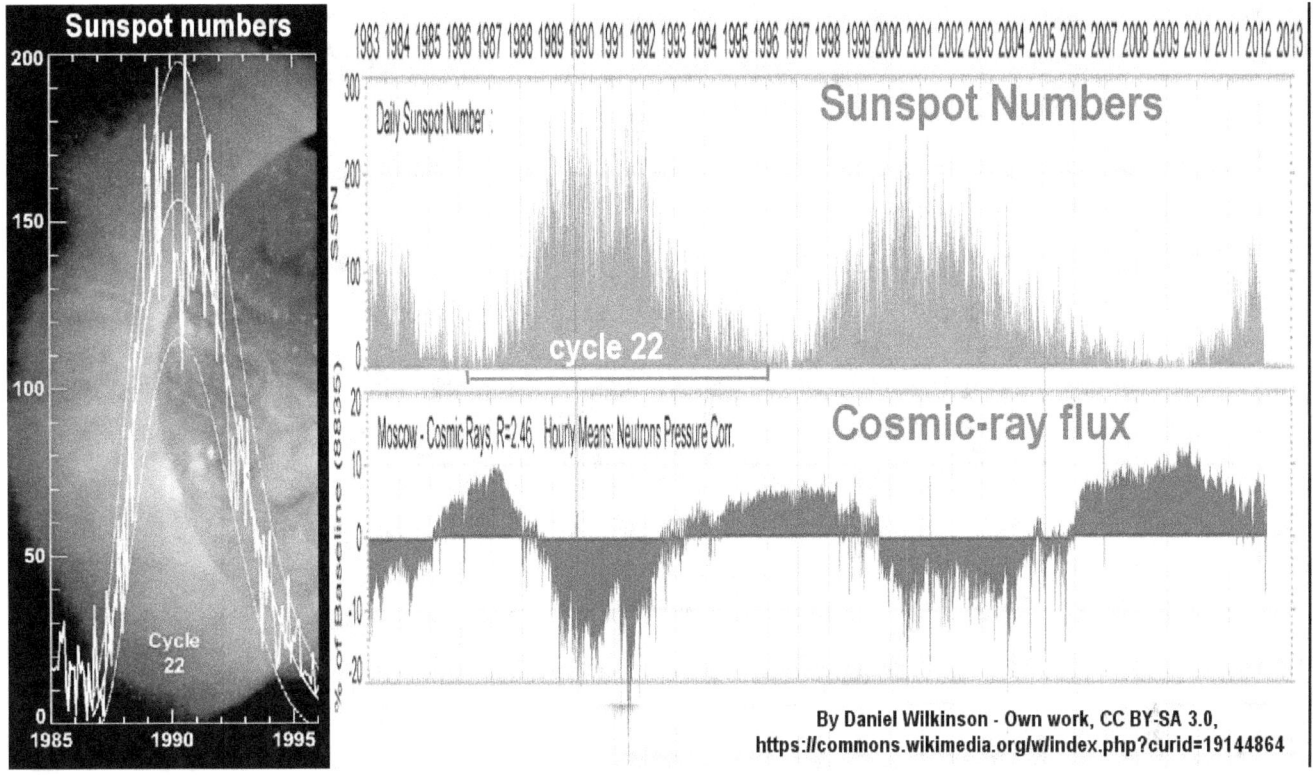

The reason why the solar activity peaks at the middle of a solar cycle is, that it takes 5 years for plasma-pressure wave from the primer fields, that is generated as a type of shock wave with the polarity reversal, to reach to the Sun. The delay is due to the long transition time crossing 7 billion kilometers of space from the primer fields to the Sun.

The plasma pressure around the Sun is a critical variable factor for solar activity. When the plasma density is high, two things happen. The solar activity becomes intense and large numbers of sunspots are created. High plasma density on the Sun also blocks to a larger degree cosmic-ray emissions from the Sun. The effect makes the peak of the solar cycle, with the densest plasma sphere around the Sun, the weakest in solar cosmic-ray flux. Inversely, it makes the valley between the solar cycles, which is weak in plasma pressure around the Sun, the period of the strongest solar cosmic-ray flux reaching the Earth, and also the period of the largest coronal holes.

The shock in the dome starts the magnetic field reversal

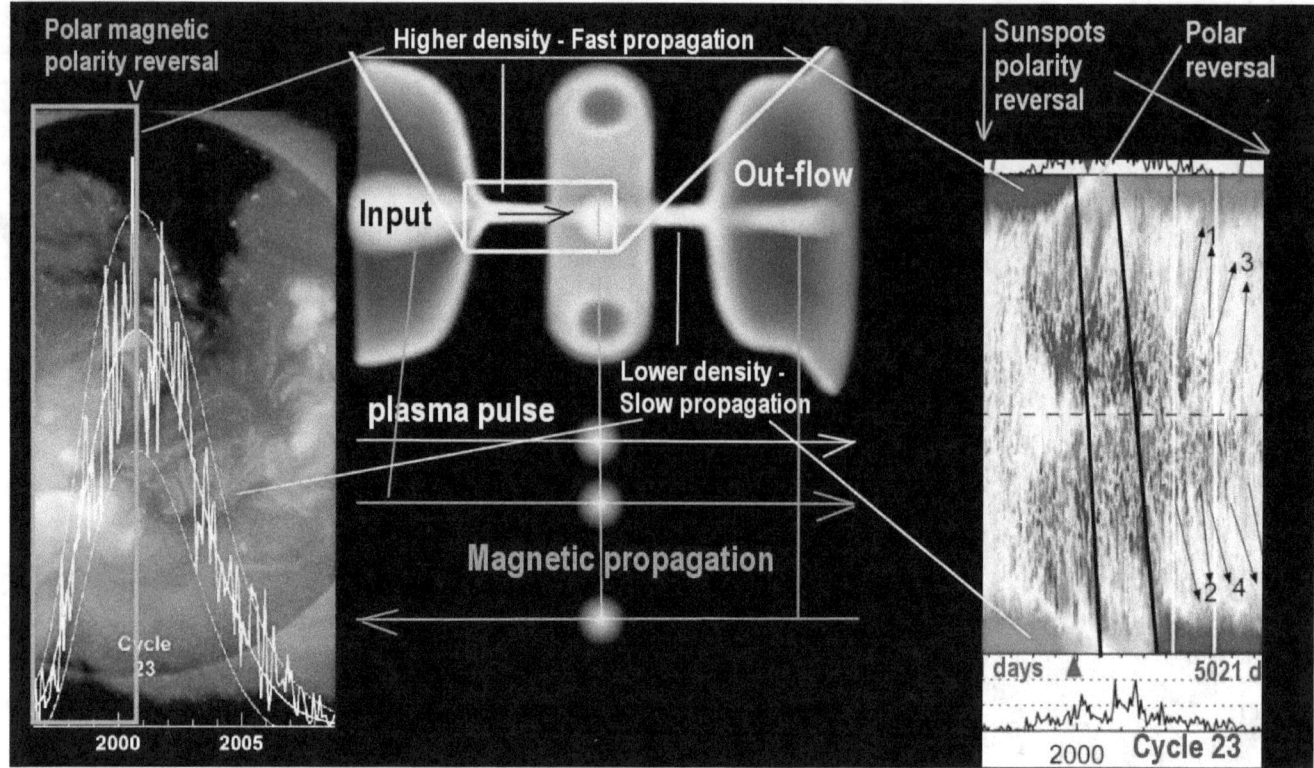

The 11-year solar cycle itself, begins with a shock event in the confinement domes. The process of breaching or closing a confinement dome's magnetic structure, changes the polarity of the magnetic field orientation that is projected from the Primer fields onto the Sun. The shock in the dome, thereby starts the magnetic field reversal, which becomes evident immediately on the surface of the Sun.

In addition, the shock event in the dome, also sets off a plasma pressure wave from the input dome. The wave builds up pressure around the Sun, which gains it maximum in typically 5 years and diminishes thereafter, as the built-up wave flows past the Sun. The pressure build-up and diminishment is reflected in the sunspot numbers. The numbers gain their maximum at maximum plasma pressure.

The shock event in the magnetic confinement dome, of course, also emits a magnetic wave along the axis of the plasma stream. The magnetic field propagation is a slow process in a high density plasma stream. But it too, reaches the Sun in about 5 years, projected from the input side. As a tightly confined stream, the projected magnetic polarity, affects only the polar region of the Sun, above the 60 degrees latitude. When it becomes active, it 'flips' the polar magnetic field orientation in the northern hemisphere.

The magnetic polarity from the out-flow side takes a little longer to arrive at the Sun, as it flows through less-dense plasma. The lower density that results, results from the Sun having consumed a portion of the plasma stream. The lower density, thereby increases the propagation time. Thus, the magnetic wave that flows from the out-flow side, affects the polar region in the southern hemisphere of the Sun, somewhat delayed.

By this differential origin, the polar magnetic field reversals do not happen at the same time in both hemispheres. The time difference can be as much as a year, or longer, reflecting differences in propagation time. But ultimately, the timing of the polar reversals are of not great significance.

The timing that separates the solar cycles from each other is not the polar magnetic reversal, but is the shock event that starts the magnetic reversal at home base, in the confinement domes.

Several types of oscillations have the potential to cause the big solar cycles oscillation to happen? The oscillation can potentially be caused by the plasma pressure wave arriving at the dome on the out-flow side. The wave arrives there after its 11-year propagation crossing 15 billion kilometers of space, flowing in a tightly confined plasma stream.

The primer fields being magnetically connected

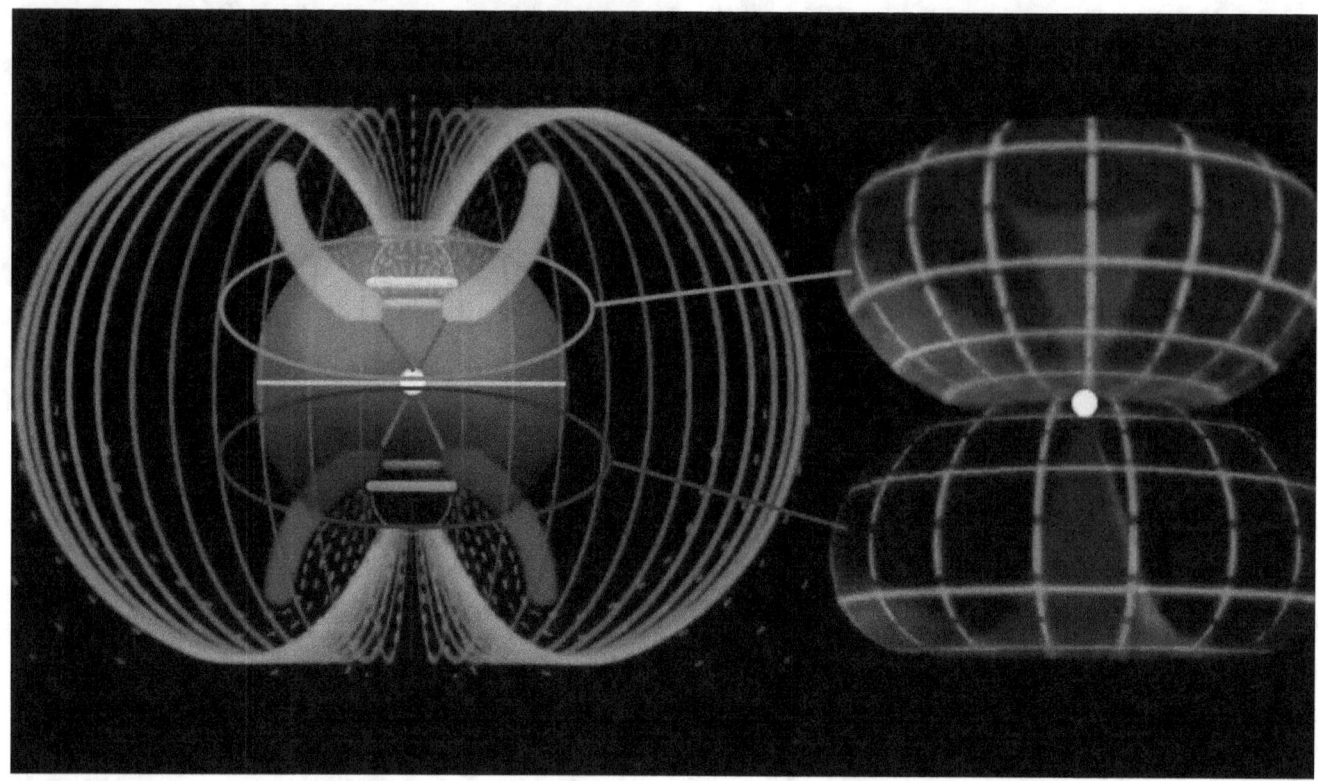

With both of the primer fields being magnetically connected with each other across 'empty' space, the pair of primer fields have the potential to be magnetically synchronized within a couple of hours. In cosmic terms, that's instantaneous.

A 2-way magnetic oscillation cycle

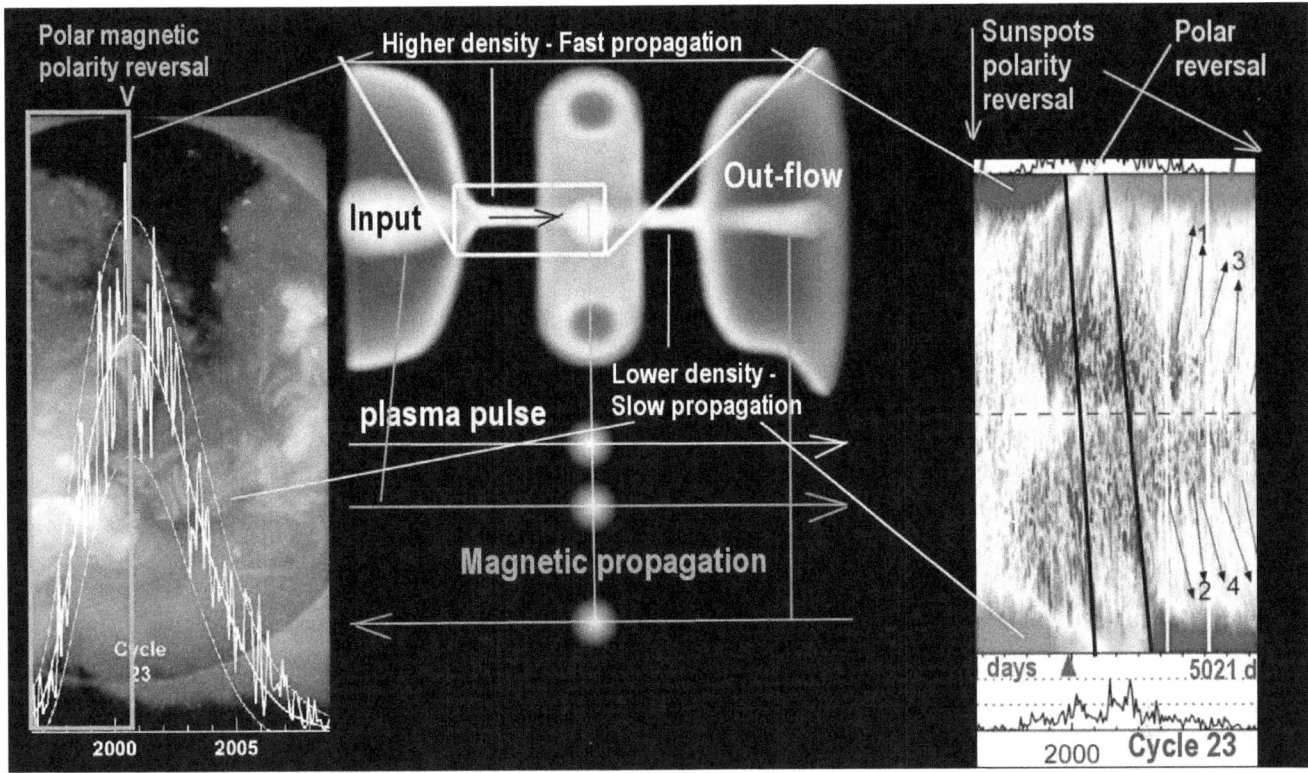

The 11-year solar cycle may also be the result of a 2-way magnetic oscillation cycle, each flowing within in the high-density plasma stream to their opposite primer fields. Each of the magnetic field projection would take the same path and reach the other's magnetic dome at the same time. This could cause a magnetic reversal to happen without external synchronization.

It could also be that both, the magnetic waves and the plasma wave, acting together, may cause the polarity reversal in the domes that starts the next solar cycle.

The solar system as a whole is diminishing

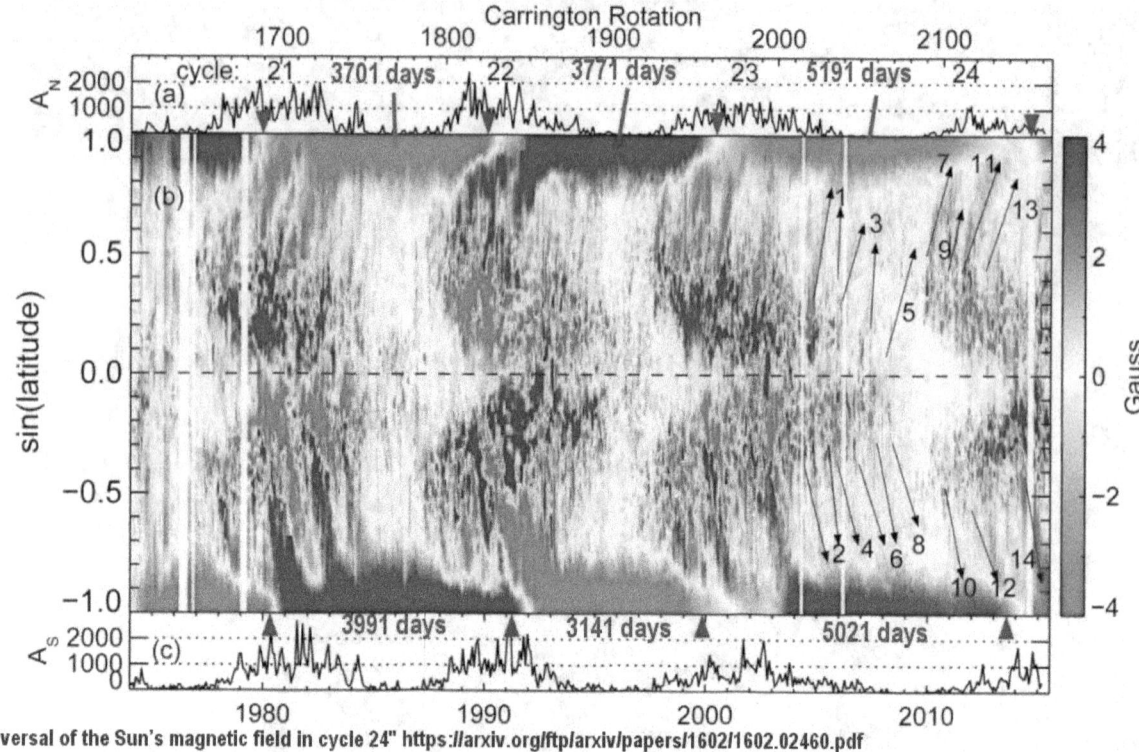

"The reversal of the Sun's magnetic field in cycle 24" https://arxiv.org/ftp/arxiv/papers/1602/1602.02460.pdf

It is worth noting at this point that as the solar system as a whole is diminishing. It has been weakening from the late 1990s onward. As the result of the general weakening, the polar magnetic field strength of the Sun has diminished accordingly. The same wakening is also expressed in slower propagation times throughout the system. This has lengthen the interval between the polarity reversals that start the next solar cycle. The lengthened interval has lengthened the solar cycle as a whole.

Something big is in progress here that has sparked significant concerns. The magneto graphic that is presented here is a composite result of a major study.

A 19-page research report

The reversal of the Sun's magnetic field in cycle 24

Institute of Solar-Terrestrial Physics,
Russian Academy of Sciences, Irkutsk, Russian Federation.
 - Alexander V. Mordvinov

National Solar Observatory, Sunspot, New Mexico 88349, USA.
 - Alexei A. Pevtsov

National Solar Observatory, Tucson, Arizona, USA.
 - Luca Bertello,
 - Gordon J.D. Petrie

https://arxiv.org/ftp/arxiv/papers/1602/1602.02460.pdf

Analysis of synoptic data from the Vector Stokes Magnetograph (VSM) of the Synoptic Optical Long-term Investigations of the Sun (SOLIS) and the NASA/NSO Spectromagnetograph (SPM) at the NSO/Kitt Peak Vacuum Telescope facility

The graphic is contained within a 19-page research report "The reversal of the Sun's magnetic field in cycle 24", by contributors from three major institutions, and data from leading-edge facilities.

Increase of the duration of the solar cycle 24

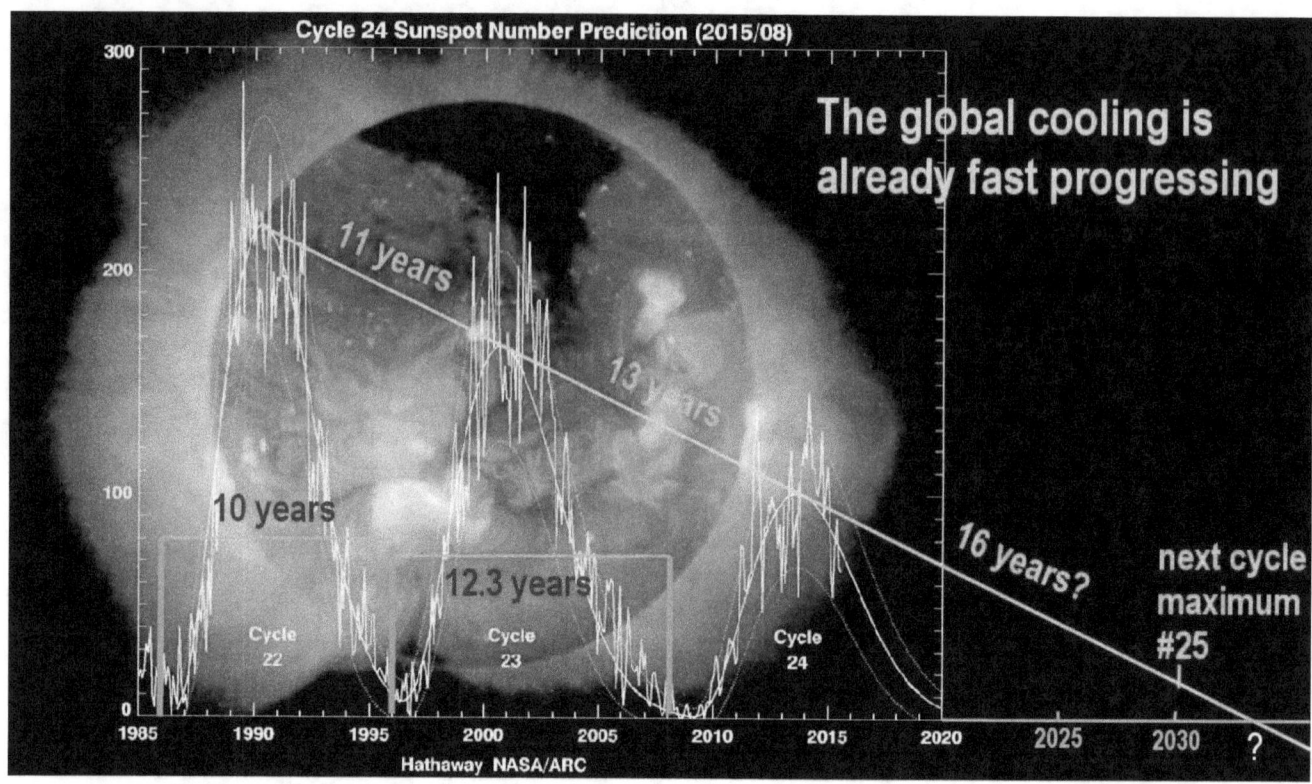

The potential increase of the duration of the solar cycle 24, is not trivial. For cycle 23, the increase in duration was 23%. Cycle 24 promises to increase that.

The weakening of the plasma streams in the solar system is also expressed in lower sunspot numbers. For cycle 23, the sunspot numbers were 15% lower. It is interesting to note too, that solar cycle 23 covered the entire time frame in which the Ulysses spacecraft had measured the solar wind-pressure, and saw it diminishing by 30%, between the start and the end of the cycle. Also note that in comparison with cycle 24, cycle 23 was still a strong cycle.

➤ The vanished northern magnetic field

> The vanished northern magnetic field

The vanished northern magnetic field.

The polar magnetic field reversal in cycle 23

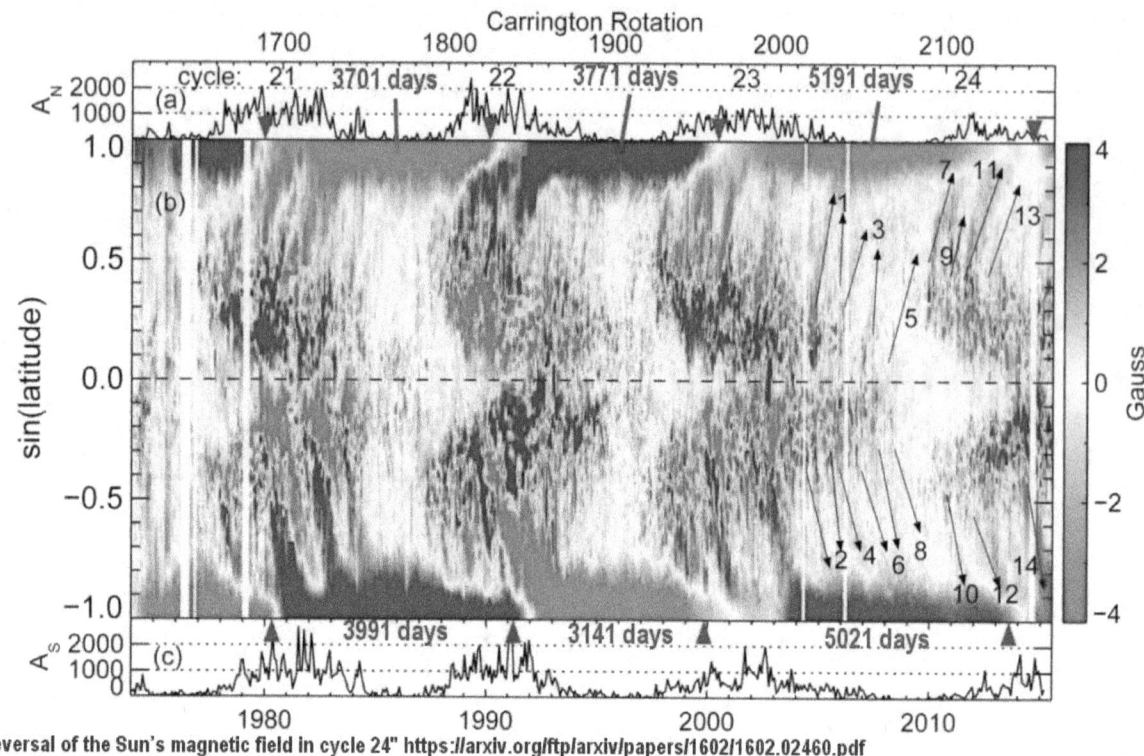

"The reversal of the Sun's magnetic field in cycle 24" https://arxiv.org/ftp/arxiv/papers/1602/1602.02460.pdf

The weakening of the plasma streams in the solar system is even expressed in the Sun's polar magnetic field strength. Note, that in comparison with cycle 22, the polar magnetic field reversal in cycle 23 was significantly weaker, and that for cycle 24, the northern reversal is missing altogether.

Polar polarity reversal didn't happen

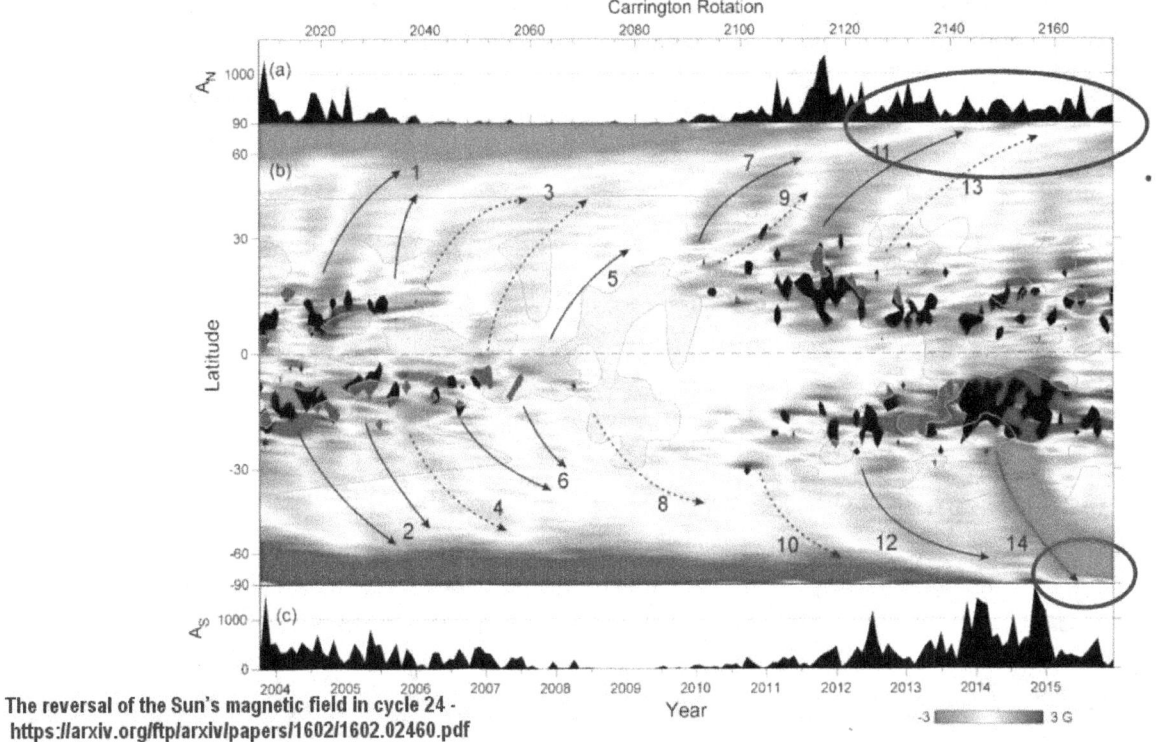

The reversal of the Sun's magnetic field in cycle 24 -
https://arxiv.org/ftp/arxiv/papers/1602/1602.02460.pdf

Solar cycle 24 began in 2008. The traditional polar polarity reversal should have happened in 2013, five years into the cycle. But it didn't happen.

Some form of reversal appears to have happened in the southern polar region, in 2015, seven years into the solar cycle, and two years late. But in the northern polar region of the Sun, no sign has been seen of anything decisive happening. up to 2016, which is the limit of the data available.

The missing northern polar polarity

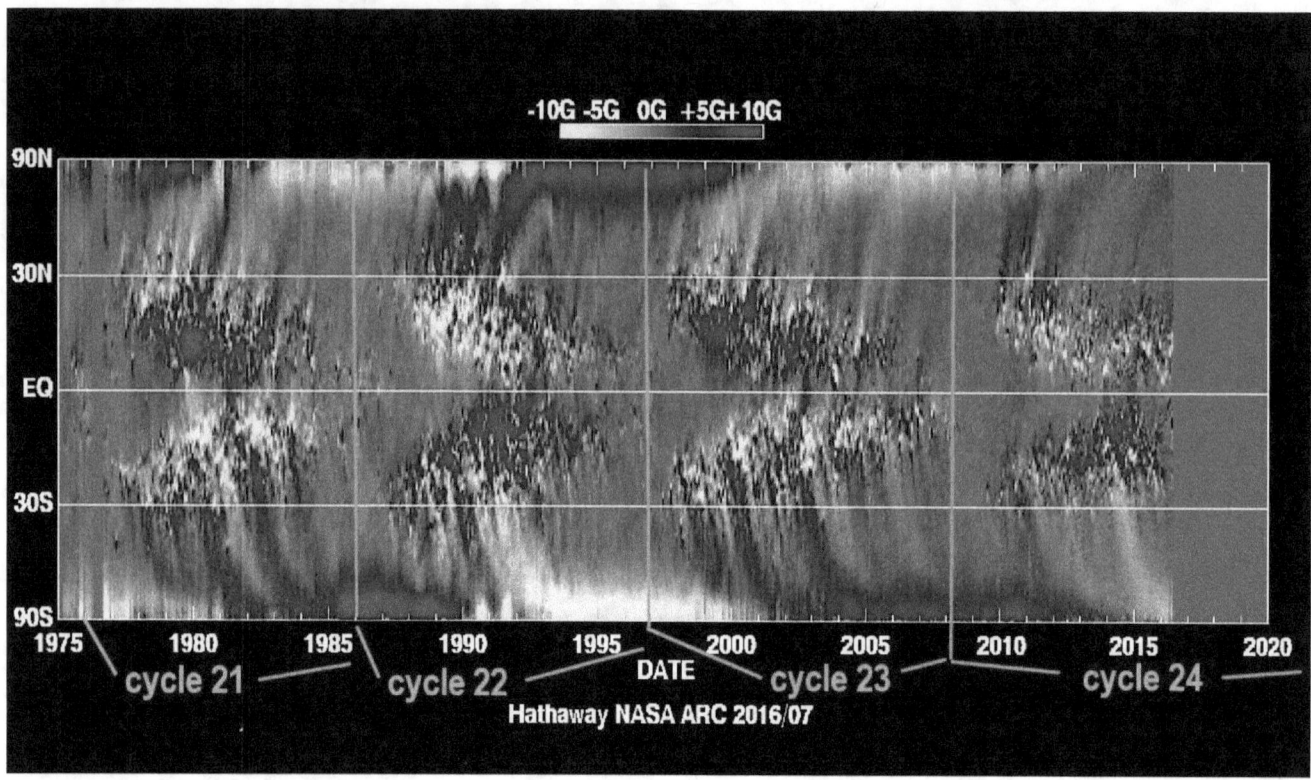

The missing northern polar polarity is also shockingly evident in the Hathaway solar magnetic graphic that covers a slightly longer timeframe. The northern solar polarity change that should have happened in 2013, isn't late this time. It didn't materialize. It didn't happen at all.

The solar system is crashing

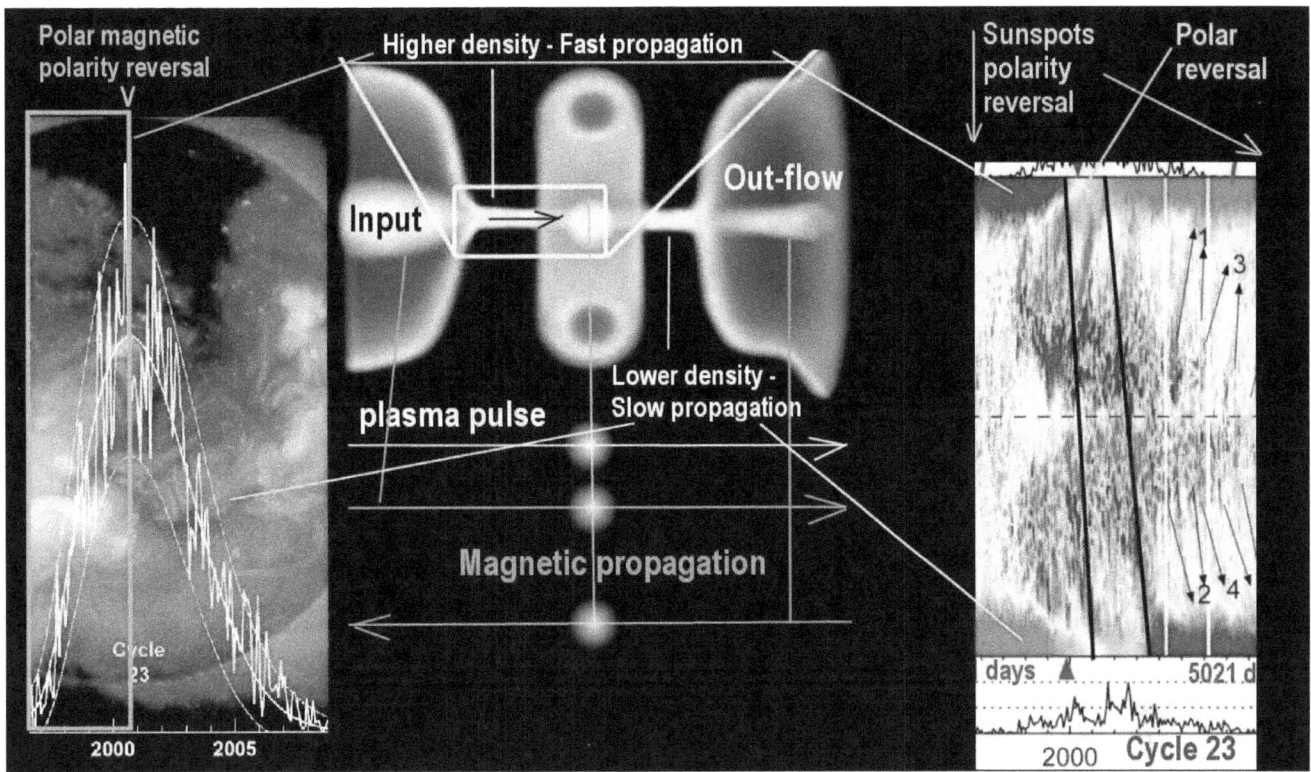

The missing polarity seems to indicate that the plasma inflow into the primer fields system was too weak for anything significant to happen. The reason why the southern polarity start-up did happen, may be due to the southern magnetic polarity being reflective. It reflects back the effect of the magnetic reversal carried by the old plasma density from the previous solar cycle, which apparently was stronger than the new density flowing into the system, which has become so weak that it no longer makes the grade.

The northern solar reversal that we saw in cycle 23, may have been the last one for the next 90,000 years till the next interglacial period recovers the now fading solar system. It seems that we crossed the bridge to no-return already in this respect.

The Sun's polar magnetic polarity may end altogether in cycle 25, and not materialize for the next 90,000 years, potentially together with the sunspots. This seems to tell us that the solar system is in the first phases of crashing. This should raise some eyebrows, because the next system reset isn't scheduled to be happening until the year Ninety-Two Thousand and Twenty. That's a long way away.

Cycles 21 and 22, gave us a yardstick

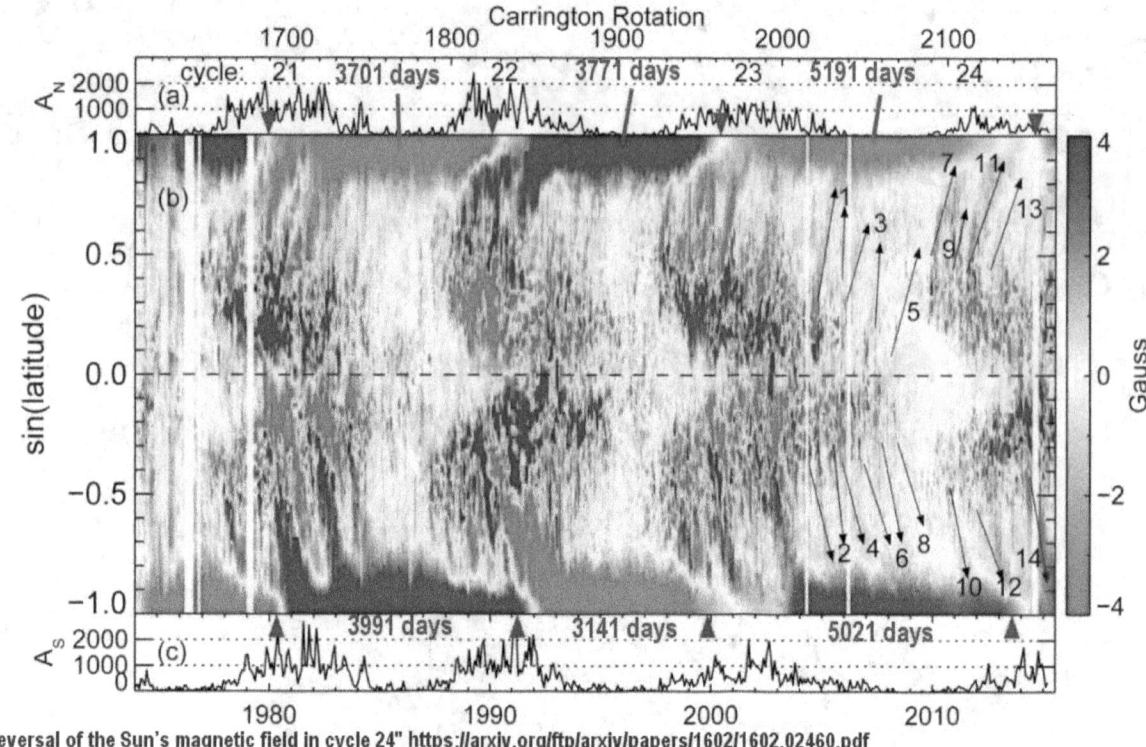

"The reversal of the Sun's magnetic field in cycle 24" https://arxiv.org/ftp/arxiv/papers/1602/1602.02460.pdf

The modern technology that we now have, enables us to measure events that we couldn't have seen prior to the 1970s. Fortunately, the early measurements that we now have available, especially those for cycles 21 and 22, gave us a yardstick with which to measure the dramatic weakening that has begun with cycle 23, and has increased in cycle 24, and is continuing to increase.

➤ When the 'shoe' drops

> When the 'shoe' drops

When the 'shoe' drops.

Solar system in a state of transition

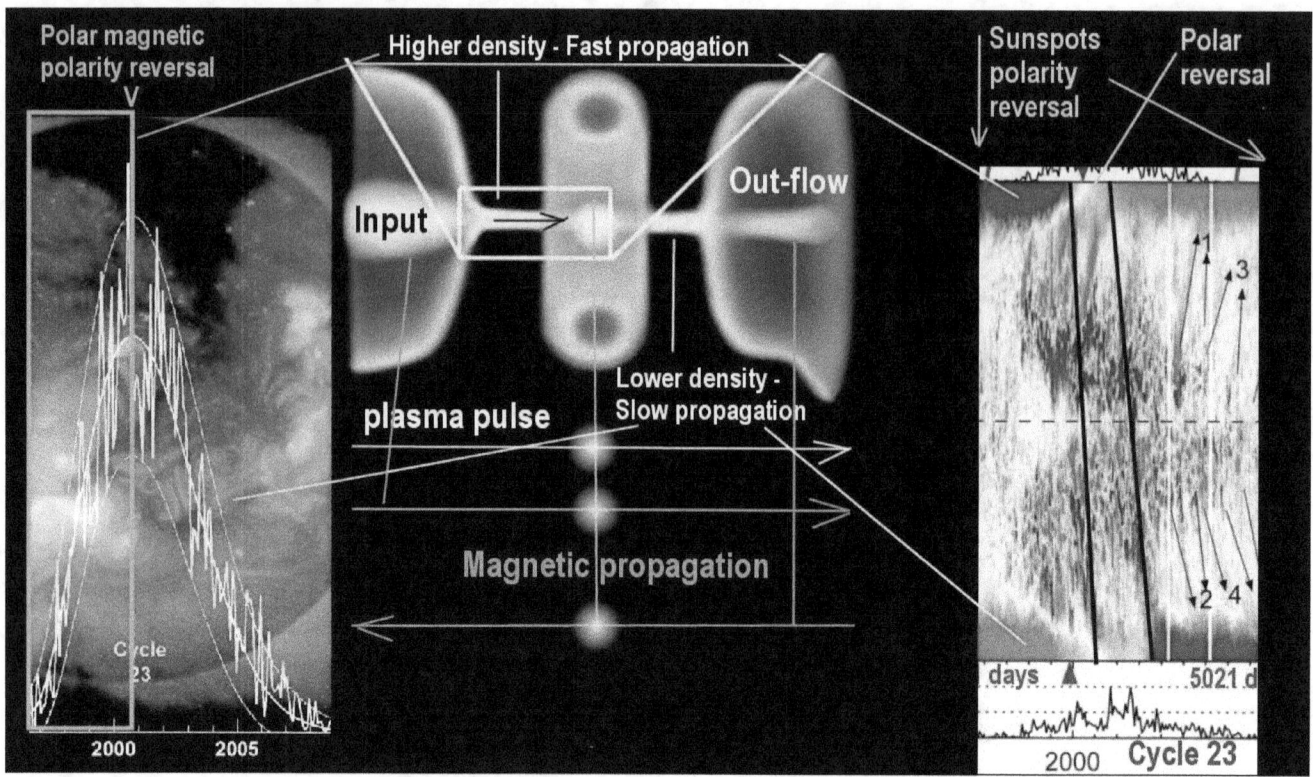

The entire solar system is presently in a state of transition. The dynamics are getting weaker, and the propagation times are getting longer. The weakening of the dynamics that we have witnessed in technological measurements, are far from being trivial.

The changes that we saw appear even more momentous when one considers that the primer fields that are critical for the solar dynamics, are themselves magnetic constructs that are formed by the movement of flowing plasma. They are formed by it. Their geometry is shaped by it. They depend on a minimal volume of plasma flowing. And that's precisely what our measurements tell us, is now diminishing, and is diminishing rapidly.

We don't know, unfortunately, where on the diminishing slope the threshold is crossed where the entire interconnected dynamic structure falls apart and vanishes as if it never existed. But we do know that when this happens, the dynamics that focus plasma onto the Sun, no longer happen. The timeframe of the high-powered Sun then ends.

A kind of hibernation state

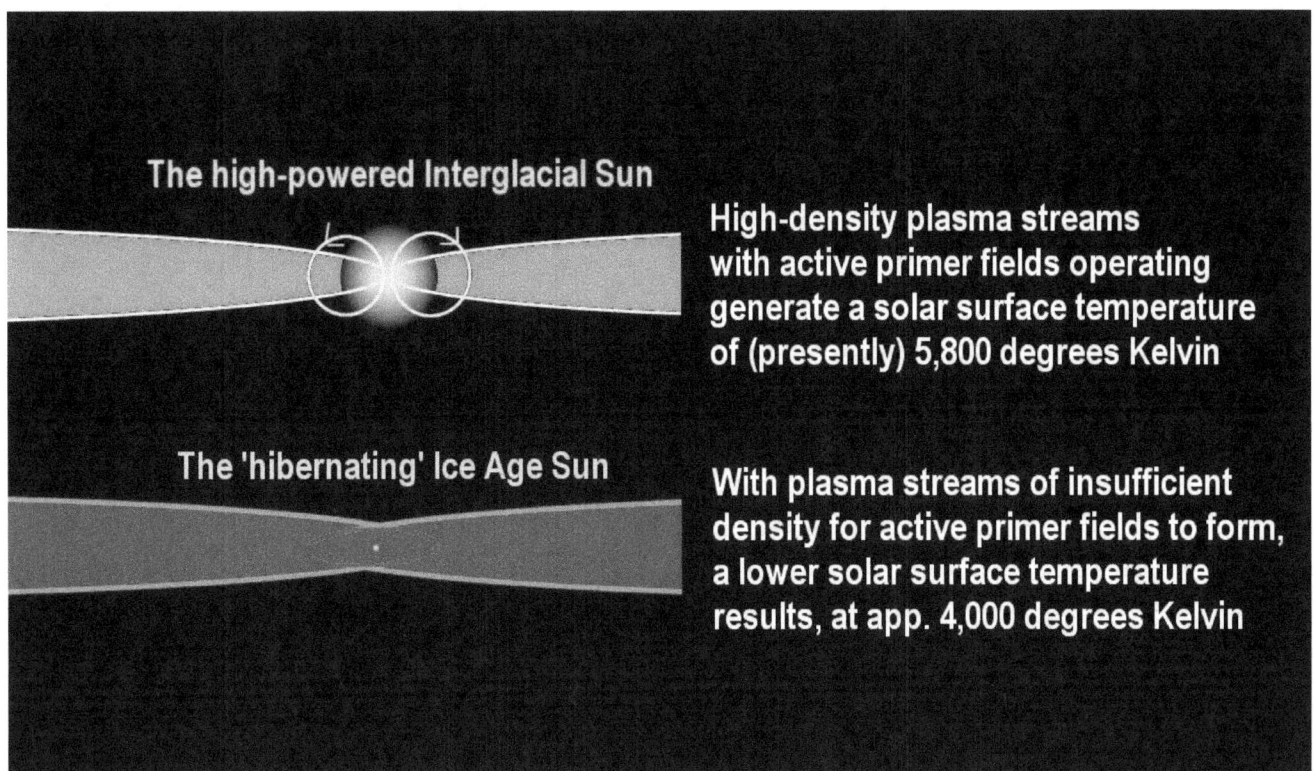

The Sun's activity level will then drop off to a much-lower-lever default state, a kind of hibernation state, where the Sun's surface temperature will be maintained at potentially the 4,000 degrees level, instead of the present 5,800 degrees Kelvin.

The resulting 70% colder Sun

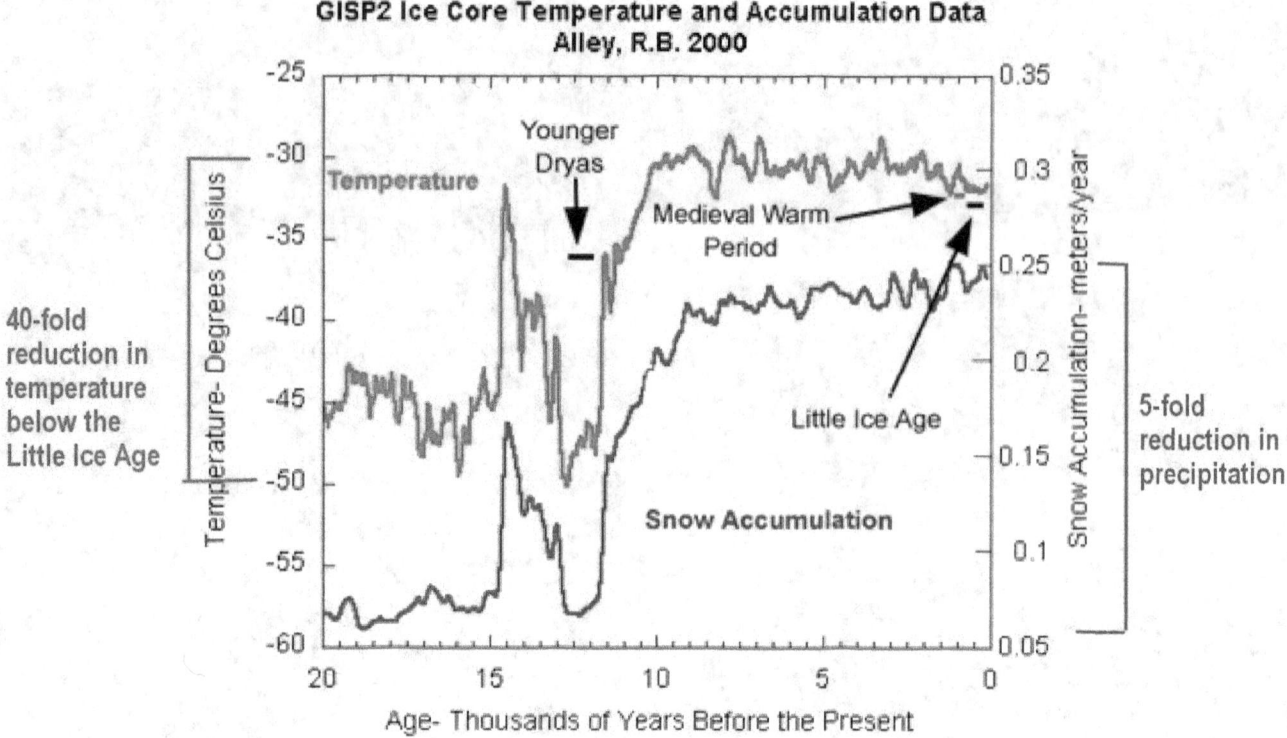

The resulting 70% colder Sun, will usher in the next Ice Age that renders the Earth largely uninhabitable, partly because of the cold, and partly because of the resulting 80% reduction of precipitation, because of the cold.

A near total loss of agriculture across the world

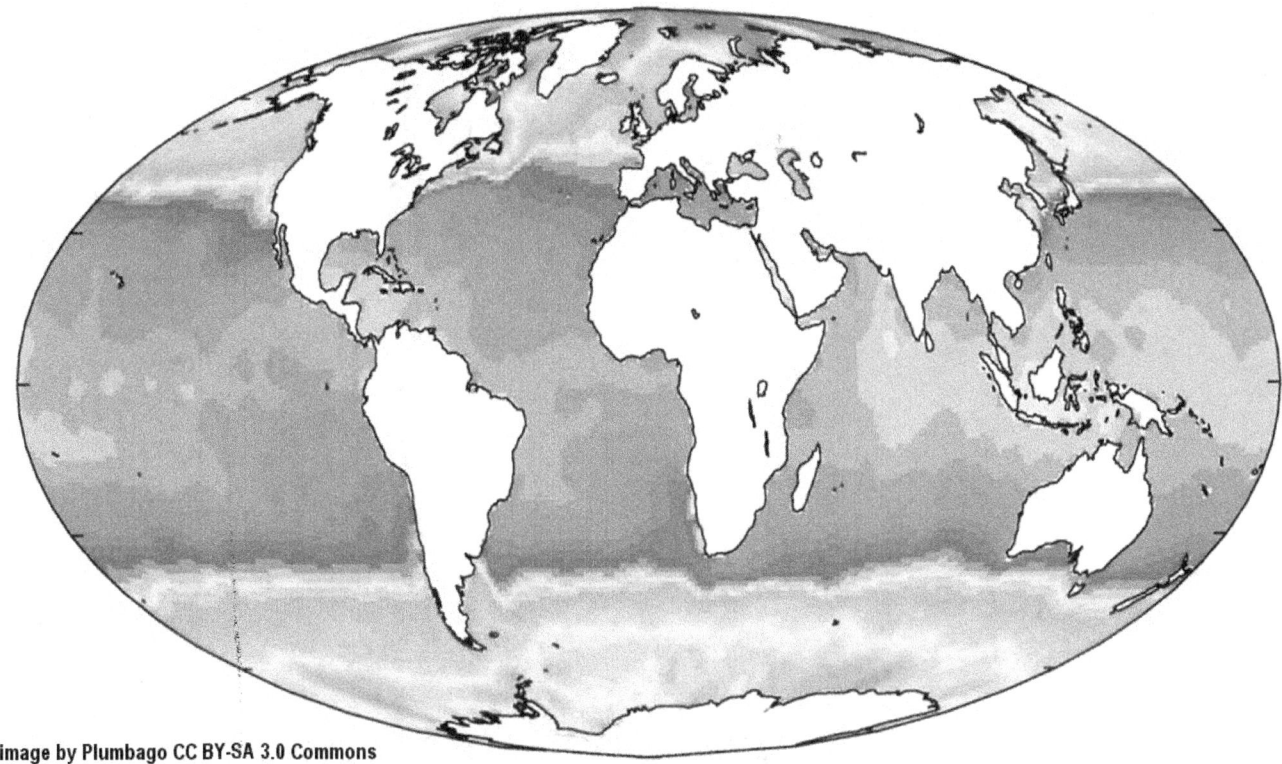

image by Plumbago CC BY-SA 3.0 Commons

Humanity cannot continue to exist under such conditions. All regions outside the subtropics become ice-bound with no hope for agriculture to continue, and in the regions that remain ice-free, the lands become deserts for the lack of rainfall. Both effects combined, assure a near total loss of agriculture across the world. That's what's at stake for the near future, potentially in the 2050s. That's what we are rushing into without anyone knowing it, and with the scientific community closing its eyes to it. Nevertheless, the consequences will be experienced with near absolute certainty. The Ice Age will happen. The process has already begun. It has been in progress for some time already.

The slowing solar heartbeat

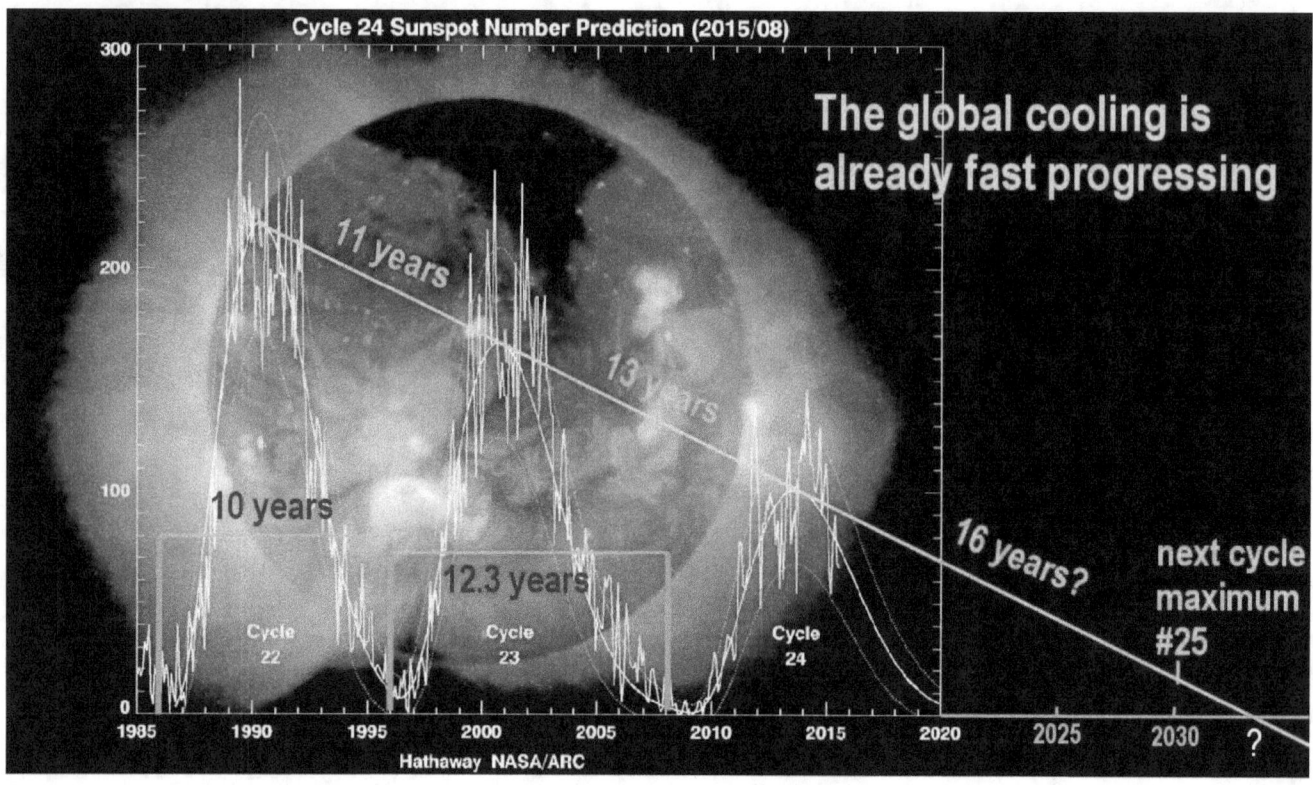

The slowing solar heartbeat, the diminishing sunspot numbers, and the weakening and missing solar polarity reversals, are merely the latest additions to the now increasing 'rush' towards the next Ice Age.

➢ Interglacial climate cooling is now accelerating

Grand Solar Minimum
becomes the Ice Age

part 2:
Uncertainty

I have pointed out in a previous video, that the interglacial climate began to cool shortly after its maximum, and that the cooling is now accelerating.

The big warming spikes began to diminish

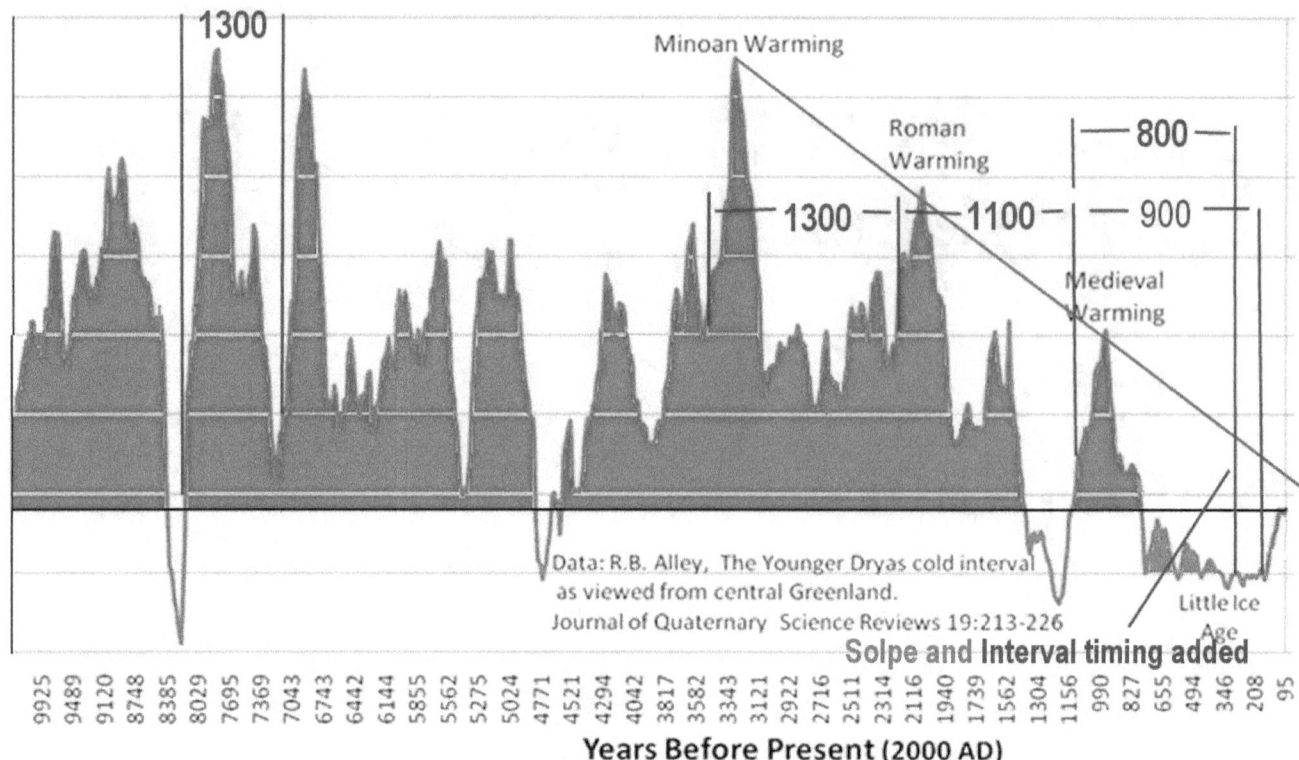

Then three thousand years ago the big warming spikes began to diminish too, and the intervals between them became shorter.

Solar minimum events became colder

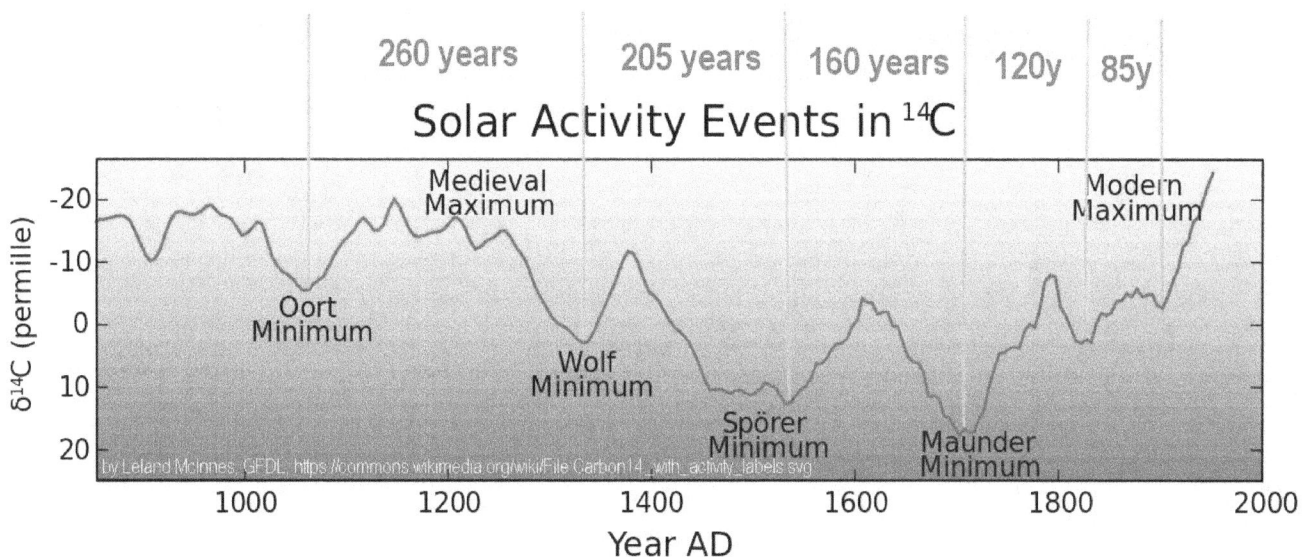

Similarly, a thousand years ago the solar minimum events became colder, and the intervals between them, likewise became of shorter duration.

While we were 'rescued' in the 1700s from collapsing back into the full Ice Age, the rescue pulse that saved us is now spent.

The solar system is rushing back

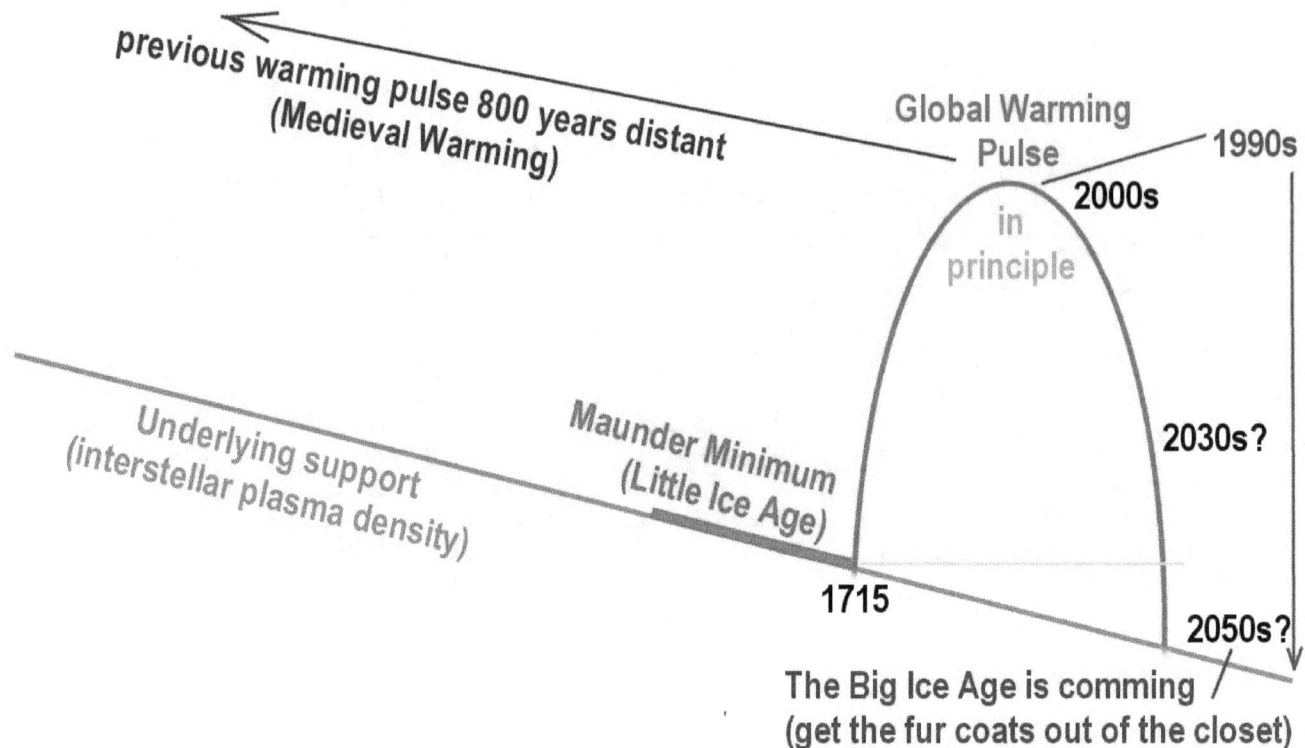

The solar system is rushing back to the conditions that we just barely escaped from, and wich are diminishing towards still lower levels.

The diminishing solar-wind pressure

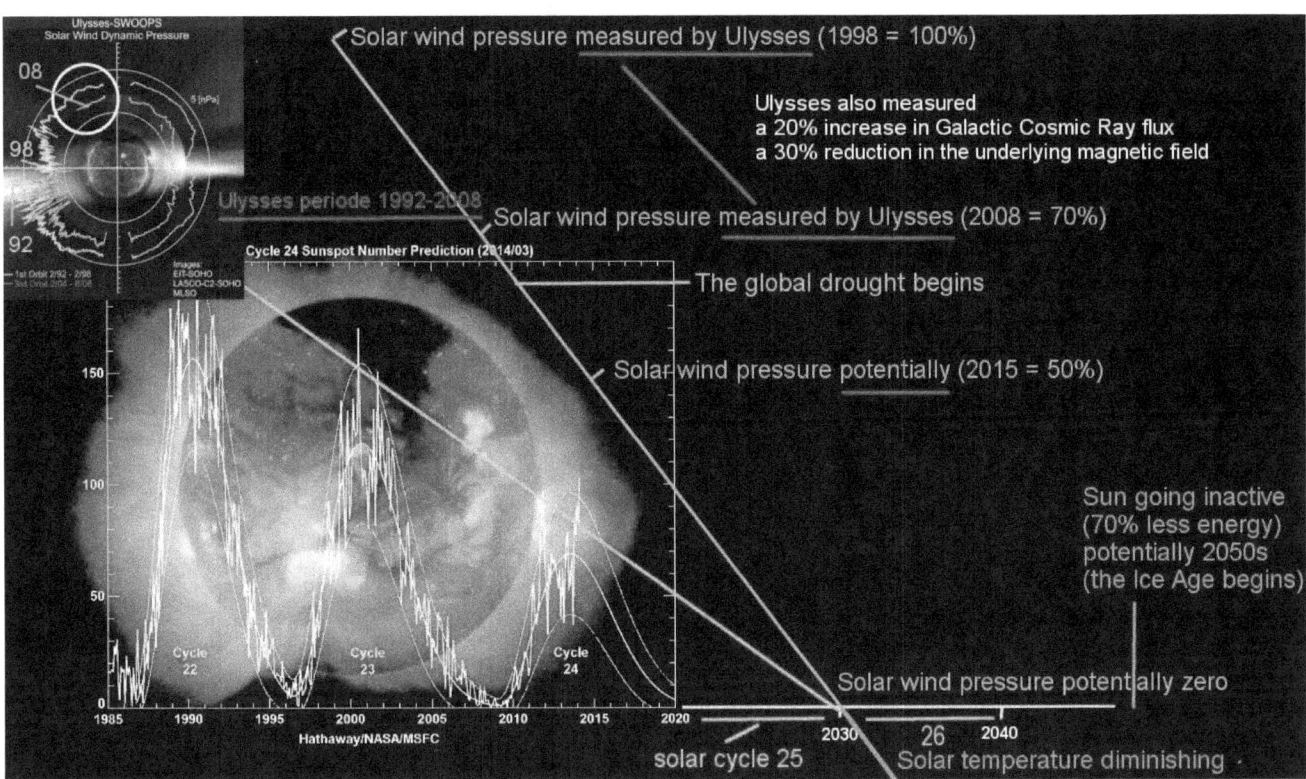

The diminishing solar-wind pressure, that began in the late 1990s, that the Ulysses satellite saw the beginning of as a 30% reduction of the solar-wind pressure in the time of a single solar cycle (cycle 23), is just another step of the ever-increasing 'rush' to the next Ice Age that we find ourselves in.

These events are real. They are not projections of something that may happen. They have already happened. They are a part of the history of the Sun. Neither are these solar events isolated events. They are way marks on a path that we cannot get away from, that is becoming evermore precarious.

Just another 'shoe' dropping off along the path

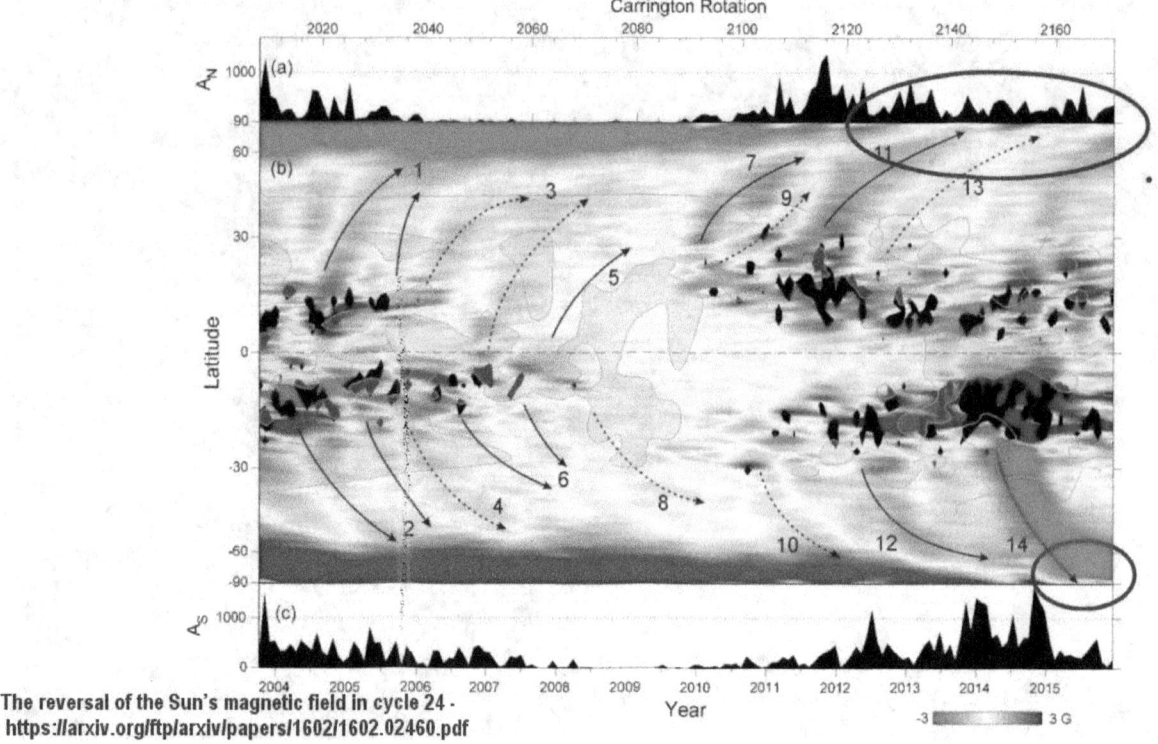

The reversal of the Sun's magnetic field in cycle 24 -
https://arxiv.org/ftp/arxiv/papers/1602/1602.02460.pdf

In this context, the now missing polar magnetic reversal of the northern region of the Sun, is just another 'shoe' dropping off along the path.

Science should be the voice of truth

As I pointed out in a previous video, the Ice Age is near, even statistically. It is scientifically understood and also scientifically hidden. Science should be the voice of truth. It may be that again some day.

The Ice Age, is a long-term 'solar cycle'

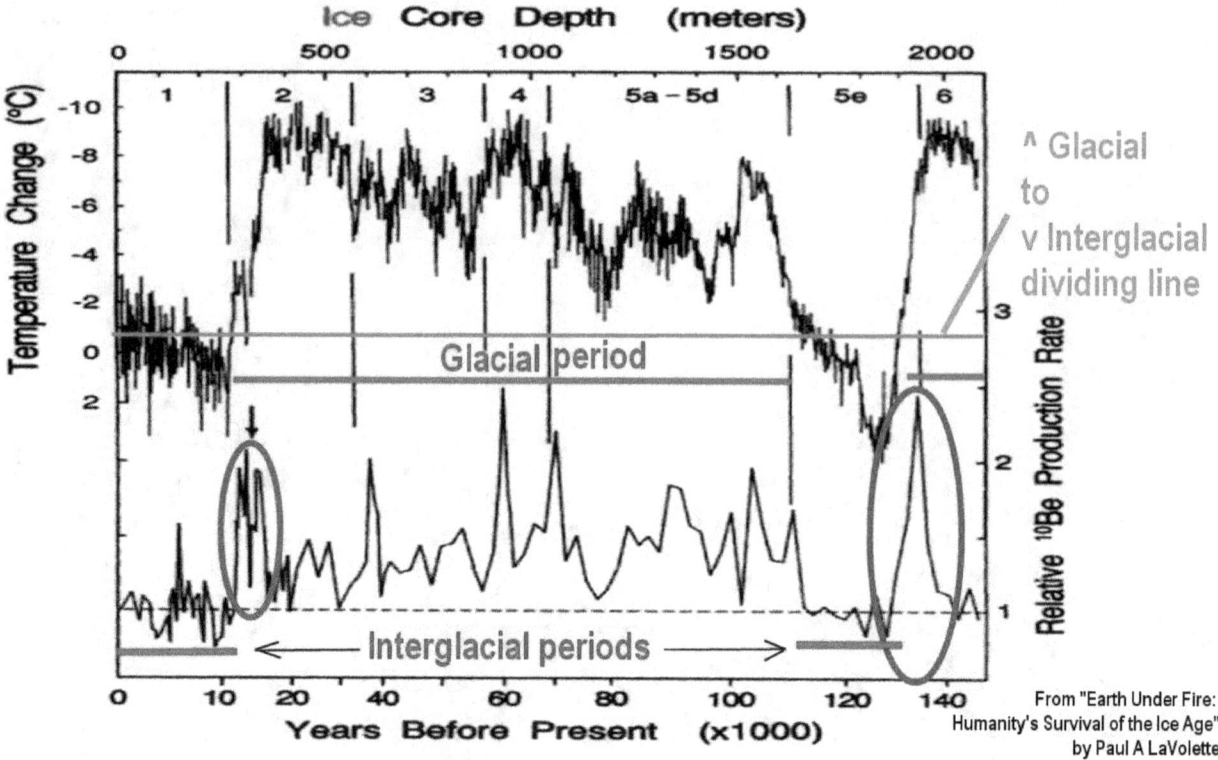

From "Earth Under Fire: Humanity's Survival of the Ice Age" by Paul A LaVolette

What we call the Ice Age, is a long-term 'solar cycle' that has been extensively researched, is extensively understood, and has been effectively buried. This tragedy needs to be reversed before it becomes an even bigger tragedy for all humanity.

The Ice Age has put us on a train to oblivion that we cannot avoid. We can only aim to sidestep the consequences of the coming Ice Age, by building us the technological infrastructures to do so, but we cannot stop the train from rolling. Nothing can stop the Ice Age dynamics, but we can act to get ready for them.

➢ The principle of universal good

$$\frac{\text{The revolutionary principle}}{\text{The principle of universal good}}$$

The revolutionary principle, The principle of universal good

To build us a new world in the tropics

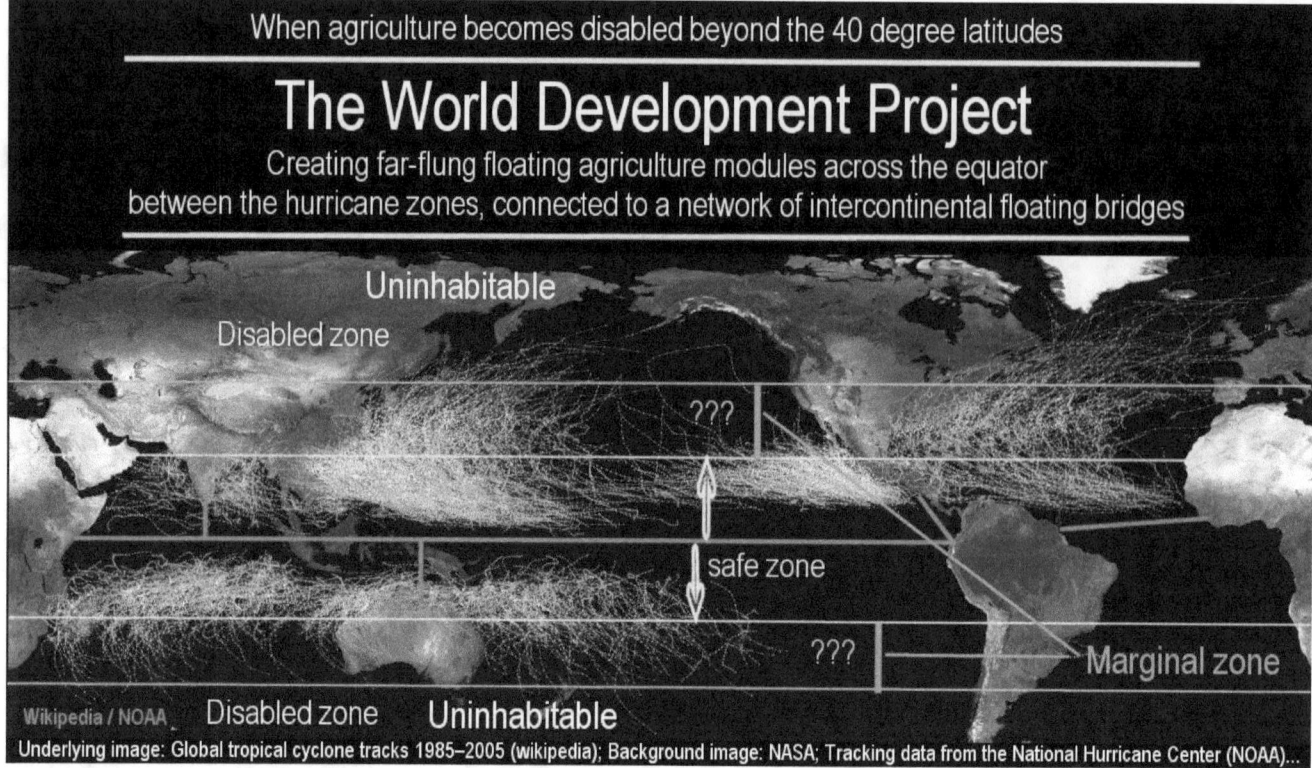

While we have the capability and the resources to build us a new world in the tropics, with modern types of technological infrastructures that the Ice Age cannot touch, nothing on this line is even considered, much less being built.

For meeting the Ice Age Challenge, we need to build 6,000 new cities for a million people each, in the tropics, preferable along the equator, with new agriculture attached to them, enough to nourish 7 billion. And since suitable land is scarce in the tropics, the new infrastructures will need to be placed afloat across the equatorial sea.

Impossible in the landscape of zero-sum economics

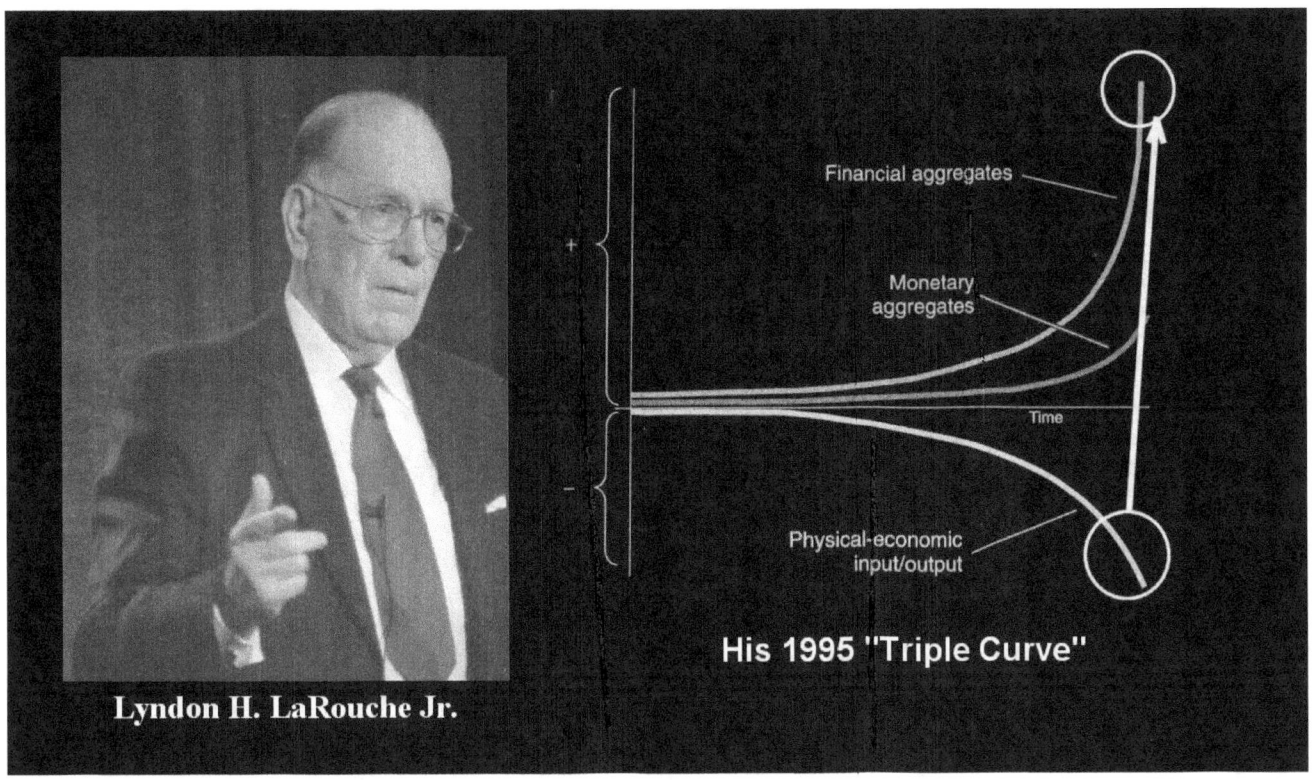

Do I here any protests here, with screams that this is impossible to do?

Of course this is impossible to do in the present landscape of zero-sum economics.

Zero-Sum equals impotence.

Losers are balanced by the winner

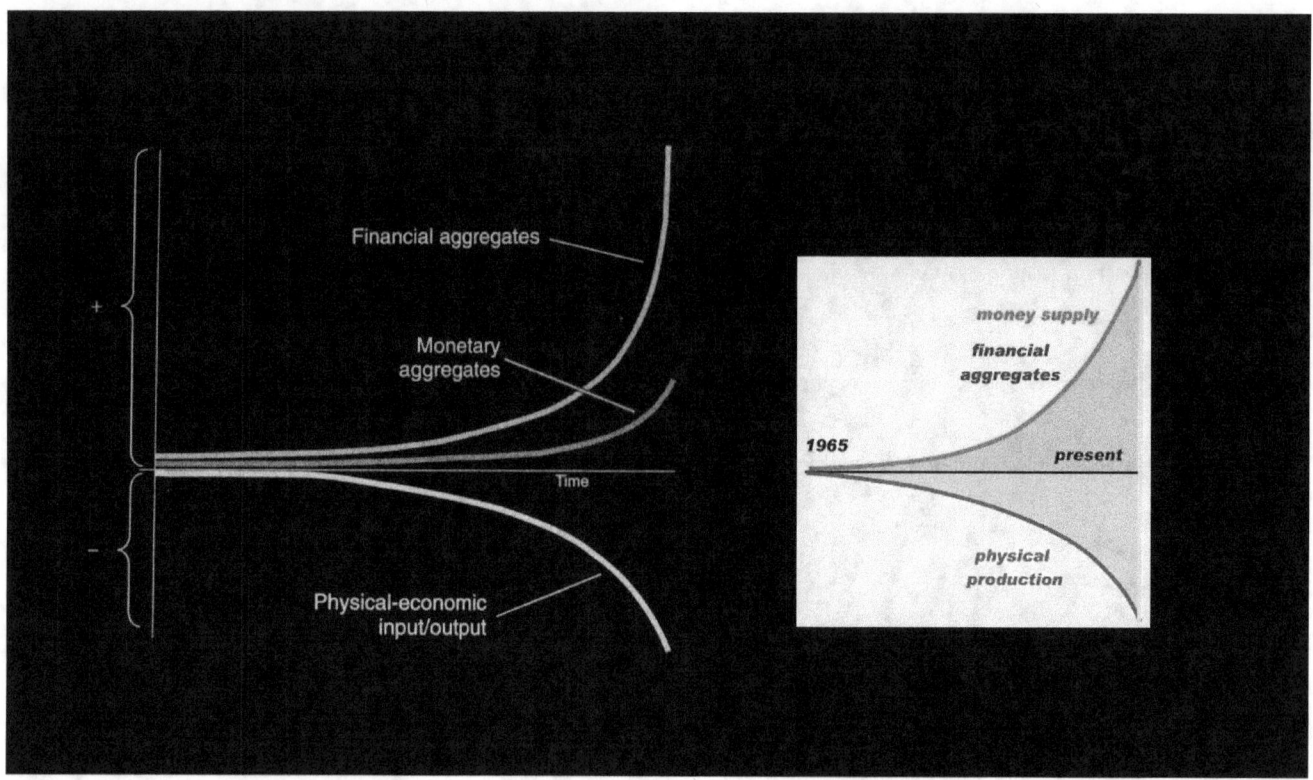

The oligarchic masters who aim to rule the world, say to society, when we rob you to the bone, the total wealth in the world hasn't changed by the process. The losers are balanced by the winner. When the differences are added up, the sum is zero.

Except this game is a deception by intention, to hide the crime. As the oligarchic masters steal liberally from society, according to the platform of liberalism, they diminish the creative and productive capacity of society, and as a consequence the world becomes poorer and poorer, till in the extreme there is no productive power left in the world, and the people perish.

The stolen wealth, of course, becomes meaningless in a world with a dead economy. When one adds this "advanced" world up, the sum-total does indeed add up to zero.

It is not possible to build anything on this zero-sum platform, much less a new world with high-end technological infrastructures.

So, what's the answer? How can we build a new world and live, and live richly?

Increase the productive capacity 100,000 fold

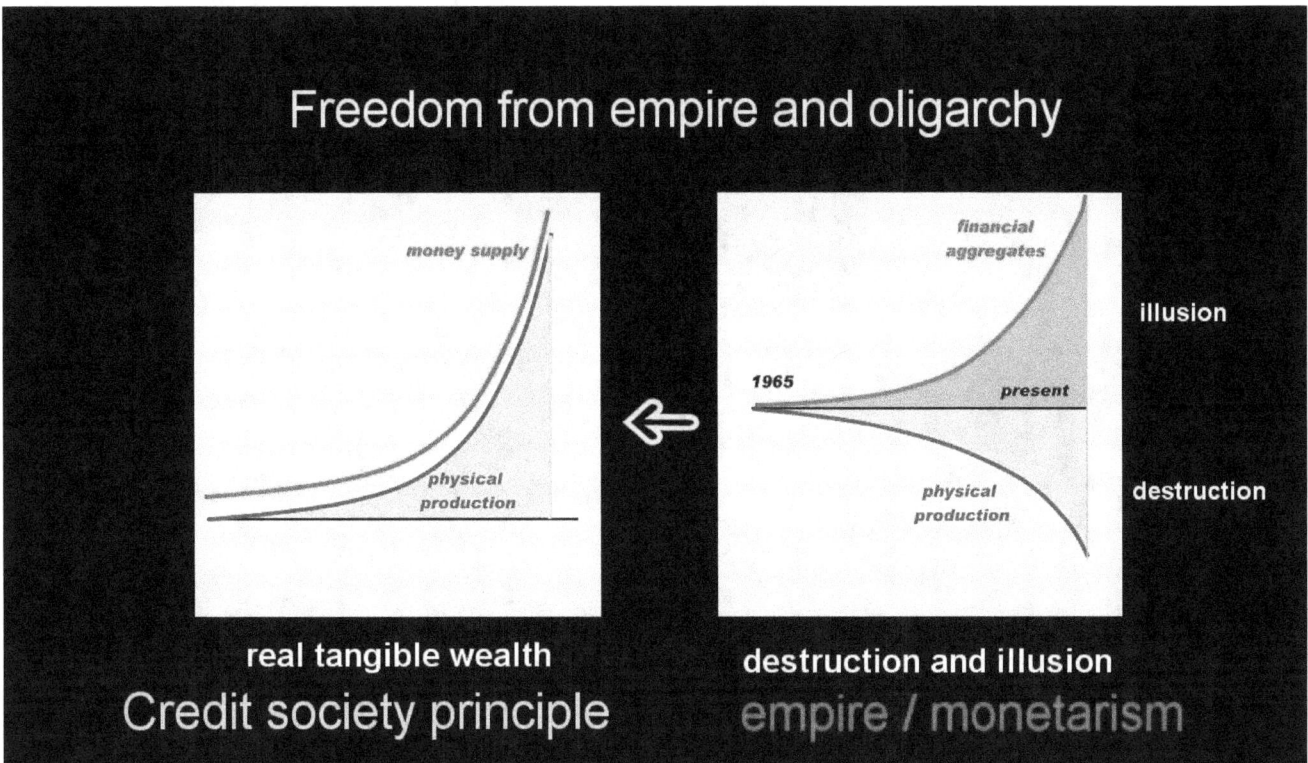

The answer lies in the opposite of the dead-end game. The answer lies building one-another up to increase the creative and productive power of society to the utmost. This means that the new cities will be built rent-free, as an investment by society into itself. No more rent slavery. Here the goal is no longer a sum of zero in the end, as in the Zero-Sum game. The goal is to increase the productive capacity of society 100,000 fold. That's the natural creative gain that is realized routinely in the universe.

An atom is 100,000-time larger than its parts

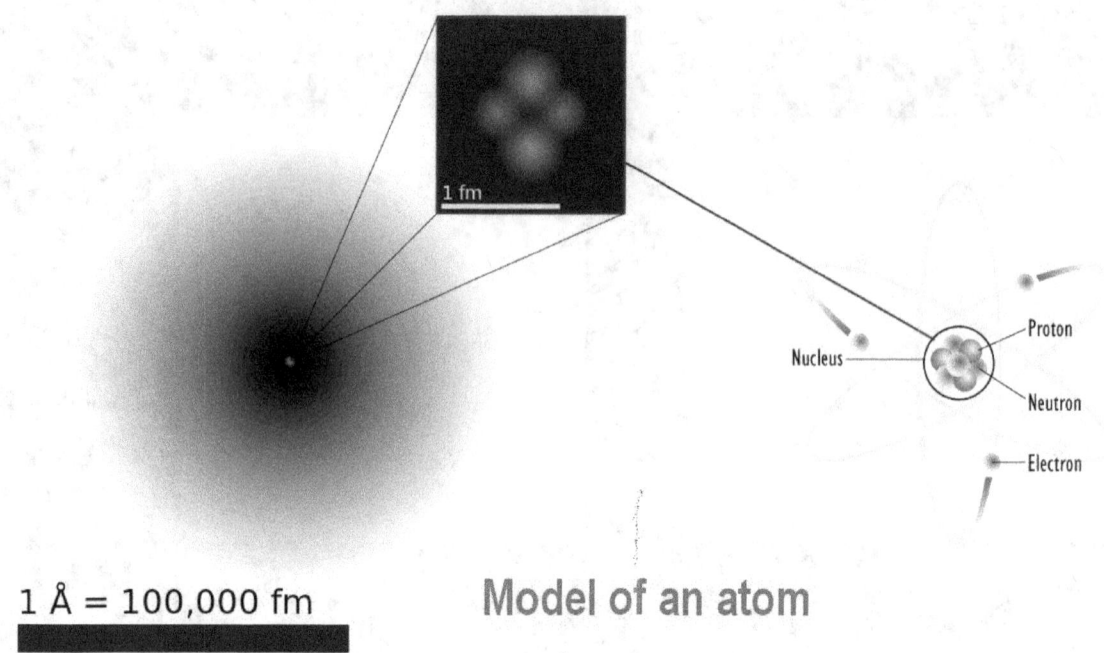

Model of an atom

wikipedia

Whenever an atom is synthesized, the resulting structure is 100,000-time larger than the sum of its parts. That's the natural dimension of creativity. On this platform anything can be built. The more is being build, the richer the world becomes.

The question will then be, what else can we build

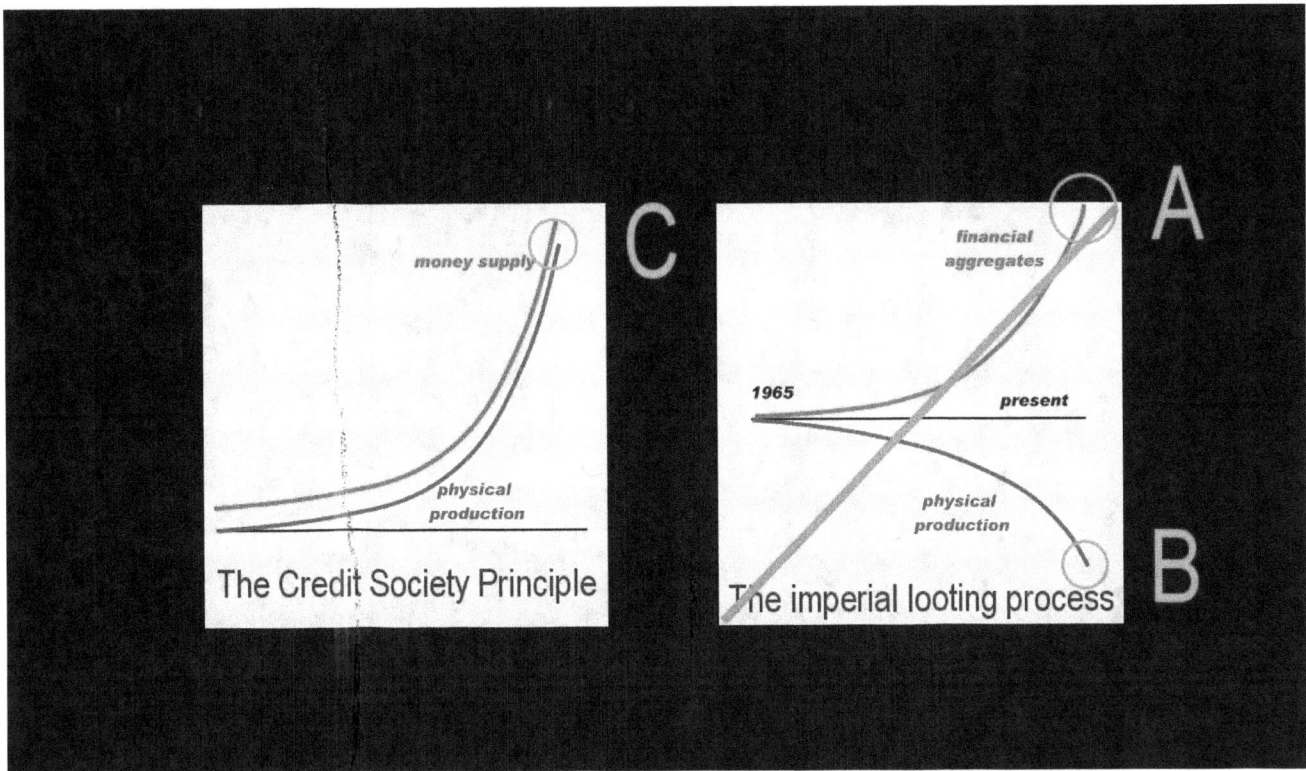

Then the foremost question will no longer be, where will the money come from? The question will then be, what else can we build to enrich our world, and our civilization within it? The money for it will be created, if money should still play a role by then.

She says 'Good' IS 'God' (universal)

A long time ago, in the late 1800s, the world-renowned American spiritual leader, named Mary Baker Eddy, had created a textbook to convey her discoveries. She had produced it with a glossary attached. In the glossary she defined the term, God. She defined it with a long list of synonymous spiritual concepts that characterize God. Surprisingly, the term, "good," is not in the list. But, isn't God, good? Isn't it the standard concept in Christianity, that God is good? How did she solve the paradox that she created, of the missing, "good," in her definition for God?

She solved the paradox in a profound manner. She defined the term, "good," separately. She defined the term "Good" AS "God." She says Good is God.

She says 'Good' IS 'God' (universal). Good is thereby not private in nature, but universal and limitless. She says in essence that if good isn't universal, there is something fundamentally lacking.

On the platform of universal good

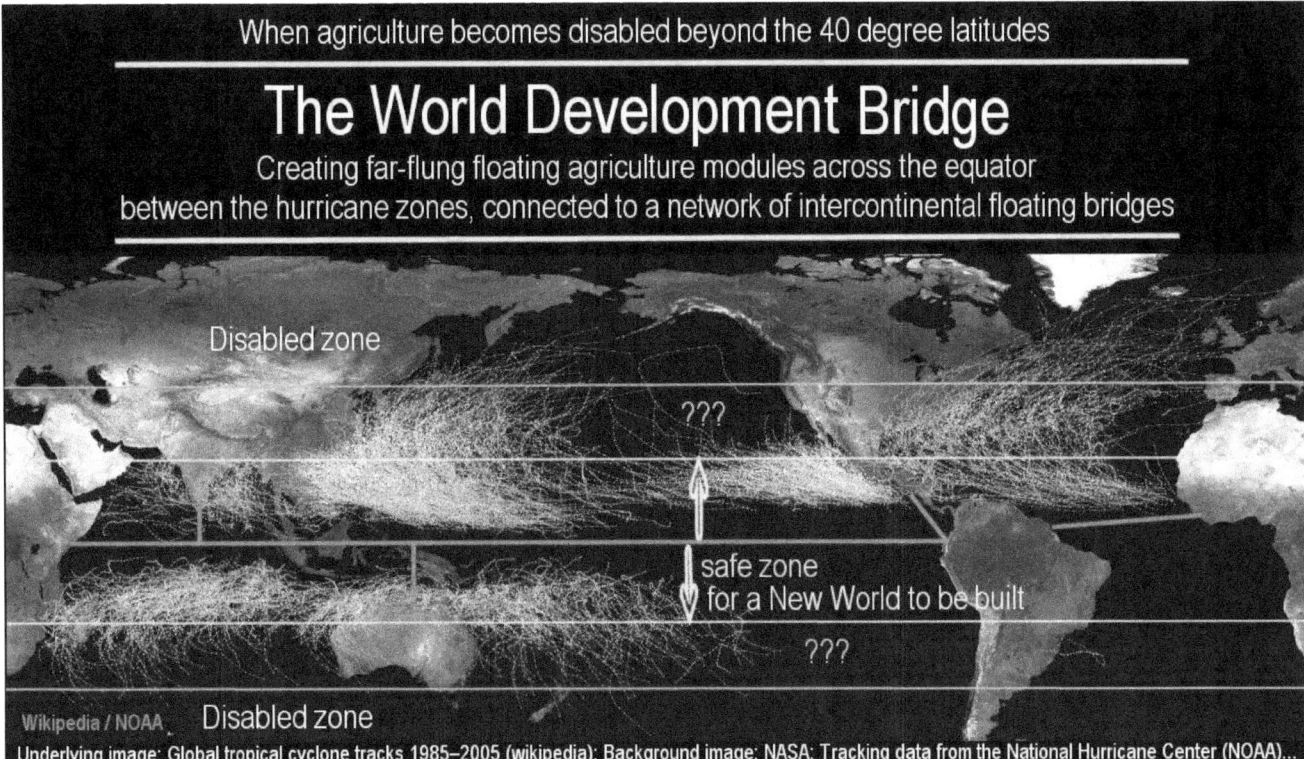

On the platform of universal good, the needed 6,000 new cities can be built, and will be built, because it fun and exciting to live and work as a human being. Universal good is the heart of our humanity.

Of course we are not at the decisive breakout yet, reaching for the platform of universal good. China has taken some courageous steps towards universal good, with the Belt and Road Initiative.

We, in the West, in comparison, presently lack the love for our own humanity to move in this direction likewise. In American history, the principle used to be called the General Welfare Principle. The principle is no longer honoured. We may claim it back, as China has done.

But these, even the best of them, are but baby steps. We need to move in leaps and bounds to match the universe and its 100,000-fold gain in creativeness. When our love for one-another will enable us to do so, then we will have a future, with a vastly richer civilization in more prosperous world than we presently have.

But at the present stage, the zero-sum stage, we cannot reach for this future for the lack of the love that would inspire us to acknowledge us as human beings, and as sons of the universe.

Some day we will see this happening, We will see us rise out of the easy chair and put the spate into the ground so to speak. Then we will have a chance.

The lack of love for our humanity is still extensive

To illustrate that the lack of love for our humanity is still extensive, is evident by the fact that we still harbour tens of thousands of nuclear bombs that we have built against each other, and keep on hand, with which to eradicate our existence - which we continuously 'improve' in terms of their destructive capability, instead of us scrapping them all. But it only takes a change in perspective to start a complete turn-around.

In the current arena of tenacious insanity

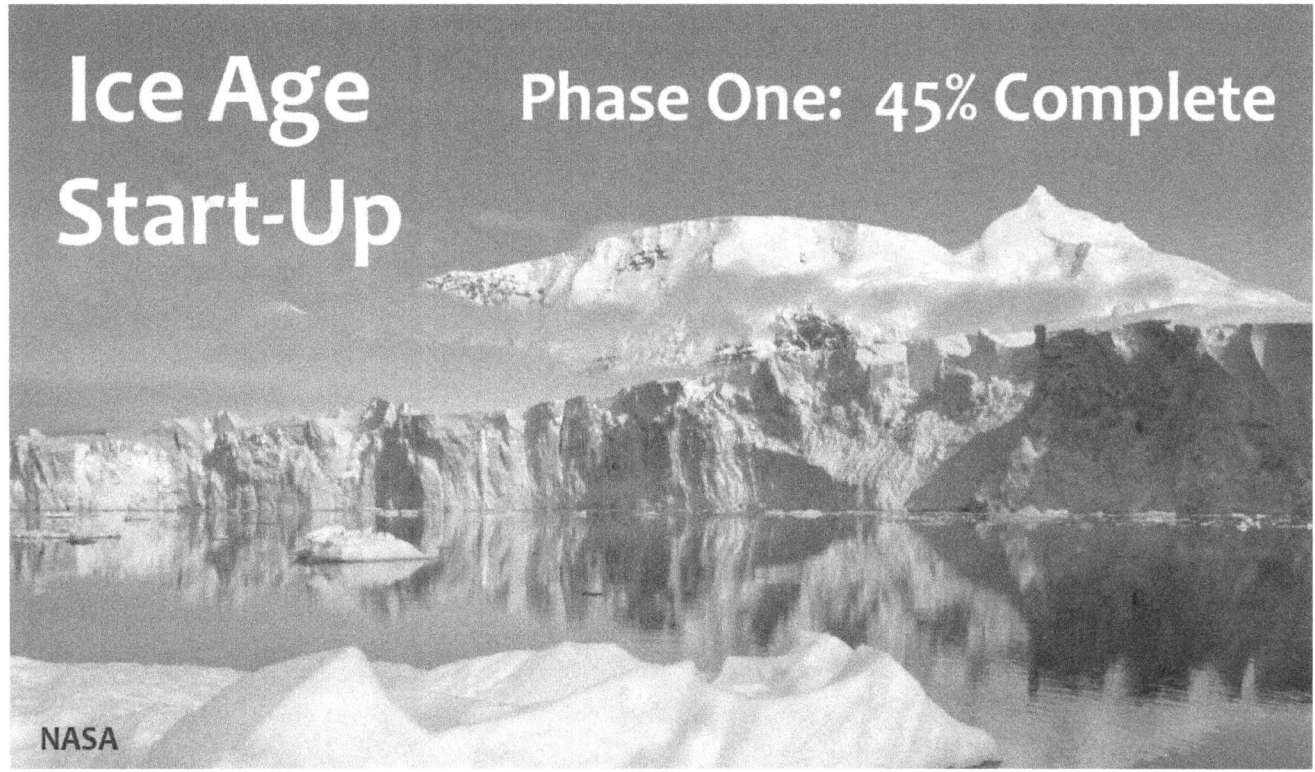

In the current arena of tenacious insanity, the Ice Age Challenge may be our final hope, as it might inspire a new focus of concern and love for our humanity that would raise us out of the nuclear-weapons trap and related issues with near absolute certainty.

In the nuclear weapons trap for 70 years

We have 'lived' in the nuclear weapons trap for 70 years, with all efforts having failed so far, for getting out of the trap.

Evidently our efforts have been too small, and our attempts too trivial.

Whereby it has the potential to change us

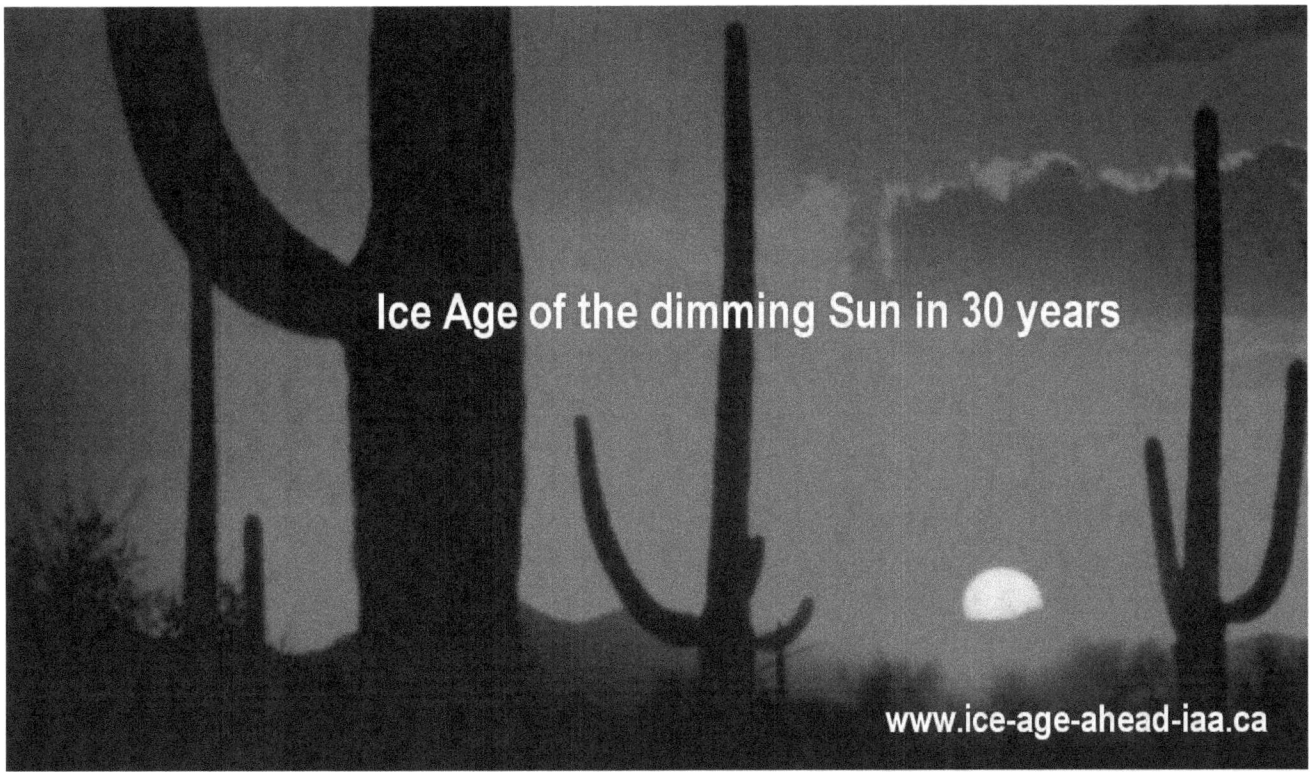

While the Ice Age phase shift is still roughly 30 years distant, its challenge is momentous and cannot be avoided, whereby it has the potential to change us and our outlook in the present, from its confrontational mode towards the principle of the universal good, meeting the common aims of humanity. If we raise ourselves up to this universal level as human beings, the challenge can be honestly, and easily addressed.

Why should we fail on this front?

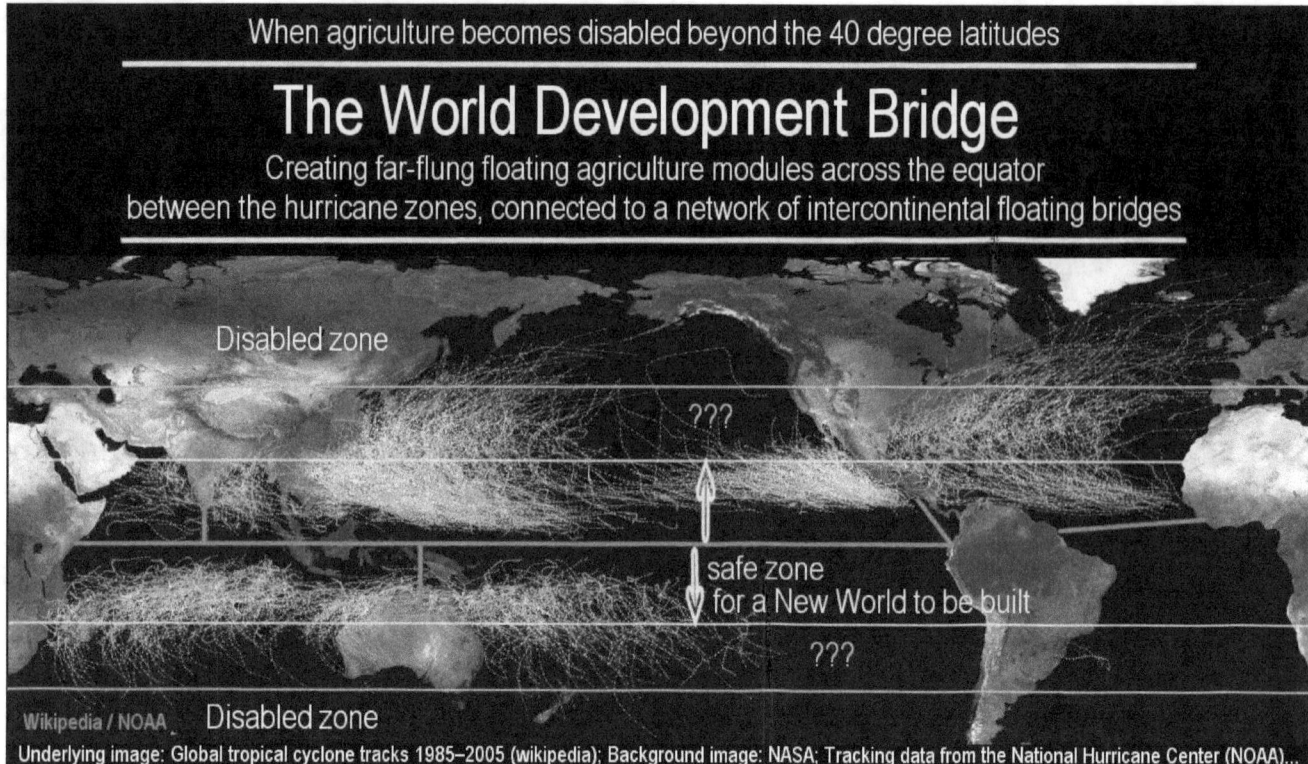

But why should we fail on this front? The prospects that we find there are tremendous. Building a new world across the sea, with new cities and new agriculture for almost 7 billion people, opening the door to greater prosperity and happiness than we have today, is imminently achievable. So why should anyone opt for death by starvation by doing nothing to prevent it, if a richer world than any yet imagined lays within our grasp? Let's reach for it and live.

   The end

➤ More from the author:

14 Libraries of books and video productions

Novels on Universal Love, the greatest principle in civilization - **14 major novels**

Flight Without Limits (science fiction)

Brighter than the Sun (nuclear war avoidance?)

A series of twelve novels: **The Lodging for the Rose**
exploring the Principle of Universal Love

Book 1 - **Discovering Love**

Book 2 - **The Ice Age Challenge**

Book 3 - **Roses at Dawn in an Ice Age World**

Book 4 - **Winning Without Victory**

Book 5 - **Seascapes and Sand**

Book 6 - **The Flat Earth Society**

Book 7 - **Glass Barriers**

Book 8 - **Coffee Sex and Biscuits**

Book 9 - **Endless Horizons**

Book 10 - **Angels of Sex in Queensland**

Book 11 - **Sword of Aquarius**

Book 12 - **Lu Mountain**

The Sex and Sacrament Project - exploration stories from my novels - **11 books**

The Son of God

Impotence and Power

Self-Love and the Golden Hijab

Erica's Flower Garden

Helen a Healer

Brilliance of a Night

Gem of the Universe

The Sound of a Bird Woke Me

Between Ice and Spirit

Anton of Grace

Goodness of Living

The Kaleidoscope Project - mixed media of stories from my novels
- videos, PDF, audio

Discovering Infinity - developing history - 13 major research books:
   A Research Book Series focused on scientific and spiritual development

   Volume ii (Introduction) **Roots in Universal History** (Focus on Reality)

   Volume 1A **The Disintegration of the World's Financial System** (Focus on Truth)

   Volume 1B **Crimes Against Humanity** (Life Denied)

   Volume 2A **Science and Christian Healing** (History as Truth)

   Volume 2B **The Lord of the Rings' Metaphors**

   Volume 3A **Universal Divine Science: Spiritual Pedagogical** (Structure for Discovery and Scientific Development - The Scientific Process to Know the Truth)

   Volume 3B **Science and Health with Key to the Scriptures in Divine Science**

   Volume 3C **Bible Lessons in Divine Science - 1898**

   Volume 3D **Living in the Sublime**

   Volume 4 **Light Piercing the Heart of Darkness** (The Demands of Truth and Justice)

Volume 5 **Scientific Government and Self-Government** (Platform for Freedom)

Volume 6A **The Infinite Nature of Man** (The Fourth Dimension of Spirit)

Volume 6B **Leadership** (The Spiritual Dimension of Leadership)

Cool Science of Kids - Illustrated Science - interactive, videos, and 20 books

War, Economics, and Nuclear War - scientific exploration - 10 videos

Civilization - series focused on humanity - 10 videos

Global Warming Doctrine - science videos - 12 videos

Freshwater and Energy - science videos - 7 videos

Christian Science explorations - 16 videos

Books by Mary Baker Eddy - Christian Science - 16 on-line books

Books by Rolf Witzsche on Christian Science - 9 Books

The Giant PDF Library all transcripts of videos in PDF form

For links, please see: http://www.ice-age-ahead-iaa.ca

The projects are designed to draw the riches of our humanity into the foreground **towards a New Renaissance**, in order that their light may out-shine the systems of empire that are erroneously accepted, including the follies of war, terror, looting, economic destruction, science-perversion, and policies for depopulation.
Rolf A.F. Witzsche

www.ingramcontent.com/pod-product-compliance
Lightning Source LLC
Chambersburg PA
CBHW062354220526
45472CB00008B/1798